WHY *Before* HOW

SINGAPORE MATH
Computation Strategies

Jana Hazekamp

Crystal Springs
SDE **BOOKS**
A division of Staff Development for Educators
Peterborough, New Hampshire

Published by Crystal Springs Books
A division of Staff Development for Educators (SDE)
10 Sharon Road, PO Box 500
Peterborough, NH 03458
1-800-321-0401
www.SDE.com/crystalsprings

Published 2011
Printed in the United States of America
19 18 17 16 15   10 9 8 7 6

ISBN: 978-1-934026-82-3

Library of Congress Cataloging-in-Publication Data

Hazekamp, Jana, 1964-
   Why before how : Singapore math computation strategies / Jana Hazekamp.
      p. cm.
   Includes bibliographical references and index.
   ISBN 978-1-934026-82-3
   1. Problem solving—Study and teaching—Singapore. 2. Word problems
(Mathematics)—Study and teaching—Singapore. 3. Mathematics—Study and
teaching—United States. I. Title.

   QA63.H29 2011
   372.7'044—dc22

                                                                2011005123

Editor: Fern Marshall Bradley
Art Director and Designer: S. Dunholter
Production Coordinator: Deborah Fredericks
Illustrator: S. Dunholter
Cover Designer: Tamara English, Bill Smith group

To my sister, Audrey, and in memory of my parents, Don and Lucille Hazekamp, for encouraging me to follow my passion for teaching. A special thank you to my dad, who passed along to me his love of teaching mathematics for understanding.

Contents

Acknowledgments

I would like to thank:

My Singapore Math family: Amy, Anni, Catherine, Char, Lorraine, and Sandy, who have been with me since the beginning on this journey of teaching math for understanding.

The many Singapore Math friends who have joined us along the way.

Fern Marshall Bradley and Sharon Smith for their extraordinary guidance through the multiple drafts and revisions that have made this book what it is. I appreciate your keen eye and friendship.

Soosen Dunholter for the drawings that bring this book to life.

The amazing and brilliant students at Pine Creek Elementary, both past and present. You are special and have so much potential. Thanks for inspiring me to be the best teacher I can be.

The talented and caring teachers and staff of Pine Creek Elementary, who work hard every day to honor all students. You are the best!

Introduction

IMAGINE BEING AN EIGHT YEAR OLD encountering a three-digit subtraction problem, 356 – 288, for the first time. Your teacher tells you to "start in the ones" and go ahead and subtract. But you can't figure out how to subtract 8 from 6. Next, your teacher tells you to cross out the number in the tens and put a little 1 next to the numeral in the ones, and then subtract. Your teacher seems to be reciting other rules to follow too, but you really don't get it. How would you feel? I'm guessing confused, just as many students are when their math instruction is based on rote memorization of rules and number facts. As a teacher, I'd want to find a better way to teach computation, and that's what I did.

Several years ago I was introduced to a series of computation strategies developed in Singapore that offer a better way to build mathematical understanding. I learned that results from the Trends in International Mathematics and Science Study (TIMSS) ranked Singapore students among the best in the world in math achievement. These results caused me to stop and look at what teachers in Singapore are doing to build mathematical understanding. I began to do some reading about Singapore Math strategies and also attended an intensive three-day workshop. A couple of key points immediately made a strong impression on me. Teachers were expected to help students build comprehension through a concrete-pictorial-abstract (C-P-A) approach. Also, they were teaching students multiple strategies for computation—not just a single rote method for each operation.

Inspired by what I had learned, I started to introduce these strategies to my students. I'm pleased to report that, as I developed confidence and skill in teaching the strategies, my students became more and more successful. What did I see in my students? One of the most significant changes was in their attitude. My students became far more engaged in their math lessons. As a group, they had an "I can solve anything" attitude that I'd seen previously in only a small group of my students. My students' improved understanding of math concepts showed when they took the initiative to apply the strategies to different and more challenging computation that I hadn't yet "taught" them. Also, they were using these strategies without my prompting when solving word problems.

What else has changed since I started teaching Singapore strategies? I notice that when my students work with partners or in small groups, they use correct mathematical language as they discuss the problems they're solving. For example, when using place value disks, my students talk about having 3 tens, or 30, to add to 5 tens, or 50, to equal 8 tens, or 80. In addition, my students are able to justify their answers with explanations as well as by sharing their reasoning when they disagree with student partners—or even with adults! And, yes, my students' test scores have also improved.

Because Singapore's methods have proved so successful in improving student achievement, more and more school systems in the United States are adopting as their core curriculum one of the several U.S. programs based on the Singapore methods. But even if your school hasn't taken this step, you can still share the benefits of Singapore Math with your students by using the material presented in this book. In fact, the school where I teach has not adopted Singapore Math as our core curriculum. I apply the strategies as a supplement to our board-adopted curriculum, and this works very well.

Ultimately, I've found that one of the greatest rewards and benefits of using Singapore strategies is that my students have become more sophisticated mathematical thinkers. Their sense of place value, ability to apply strategies on a daily basis, and willingness to play with numbers while searching for solutions and building number sense continues to show me that these strategies can and do empower my students as mathematicians.

Why Singapore Math Works

A first step toward more effective math instruction is to move away from an emphasis on teaching computation as a series of rote rules. Computation is about students comprehending what they are doing, not simply following rules. I like to make an analogy to reading instruction. In reading, the goal is not simply for students to be able to speak the words they are reading out loud. We would never say a student is a competent reader based only on his ability to "say the words." The goal is for students to comprehend and interact with text. The same is true for mathematics. It's not sufficient simply for students to calculate the answer to a problem by rote. Comprehension should also be the goal in mathematics.

For long-term learning to occur, students must first gain a solid foundation of conceptual understanding. Richard Skemp's research on relational understanding versus instrumental understanding underscores this point (Skemp, 2002). Instrumental understanding is the ability to follow rote rules—to know what to do to find an answer. Relational understanding requires students to understand both what to do and *why*. Consider the expression 3 x 4 as an example. Following an instrumental approach, a teacher would simply state the rule that multiplication is repeated addition, and thus 3 x 4 is 4 + 4 + 4 = 12. To build relational understanding, a teacher might encourage students to draw a picture or come up with a context such as 3 baskets with 4 apples in each. This would help students understand *what* is happening and *why* they can compute 4 + 4 + 4 to find the product of 3 x 4. We should aspire to build relational understanding for our students. The guided conversations in this book provide in-depth examples for computation instruction that will lead to relational understanding.

As you begin to adopt Singapore strategies into your teaching, don't forget an important point that you already know: students learn in different ways. Teachers must use multiple strategies to teach the same concept because what makes sense to one student may not make sense to another.

Also, according to Zoltan Dienes's perceptual variability principle, children will gain the most conceptual learning when they are exposed to a concept through a variety of materials and experiences (Dienes, 1971). Thus this book works to teach computation using a variety of strategies, so that we can reach all students and so that each individual student has the greatest potential to master the concepts being taught.

Following the C-P-A Approach

How can we as teachers help our students to find the meaning in the math? Research by Jerome Bruner (Bruner, 2000) states that instructional strategies build understanding for students when they move from the concrete (manipulatives) to the pictorial (visual models or drawings) to the abstract (symbols). Through the use of Singapore Math strategies, this book puts the C-P-A approach into practice. The goal is to help you develop students who not only are capable of solving computation problems, but who also truly understand computation and can apply this mathematical thinking and learning to new situations. These strategies are presented in a recommended sequence, but the strategies can also be selected and taught individually to help any student who's struggling with a particular computation concept. Also, these strategies are not just for teaching basic computation of whole numbers. They also work well for helping students master computation skills involving fractions, decimals, and units of measure.

Working with manipulatives is a first step in building mathematical understanding, but students at the early elementary levels often develop misunderstandings if they are rushed ahead without a strong foundation in base ten proportionality. To avoid this, make sure your students first work with proportional manipulatives such as base ten blocks, coffee stirrers, craft sticks, bean sticks, and straws—both as individual units and as bundles of ten—to build their understanding of proportionality (Van de Walle et al., 2009).

Once students have used proportional base ten manipulatives, they can be introduced to place value disks, which are nonproportional—their physical size isn't proportional to the quantity they represent—and thus more abstract. It is critical for kids to grasp that, although all place value disks are the same physical size, the disks represent different values of 1, 10, 100, and 1,000. When students understand that 1 and 10 are very different in value, they are ready to begin using place value disks. I recommend introducing place value strips and disks on the same day so the kids can see, for example, that a 50 strip is the same as 5 tens disks. I use them interchangeably so that my kids will be equally comfortable thinking of 50 as 50 or as 5 tens. You may decide that you will introduce place value strips on one day and disks on another day. That's okay. You must do what's most effective to build comprehension for the group of children you're working with.

Students need lots of opportunities to use many types of manipulatives as a tool for learning and understanding each type of computation. Whenever you introduce a new topic, such as subtraction, begin the C-P-A process again by starting with manipulatives. As you introduce new operations, I recommend that you not allow your students to record abstractly. It's important to keep them focused on the action so that they will understand the why as well as the how. Again, this is teaching through relational understanding, which is so important for long-term retention of concepts (Skemp, 2002).

After using manipulatives, students will also need an opportunity to use pictures as a bridge to the abstract. They do this by drawing images on paper to represent what they would be doing with manipulatives to solve a problem. Students who were working with disks would draw disks, while students who were using buttons would draw buttons, or whatever type of manipulative they'd been using. In some cases, students may need to work with the manipulatives and draw pictures simultaneously to see the relationship between the two.

Once your students have completed many examples where they work with manipulatives, you can begin to model how to record pictorially, then abstractly. This allows students to continue to focus on moving the manipulatives while also observing how these actions would be recorded on paper. As you model, talk about how the recording matches the action (you'll find this type of lesson presented in some of the guided conversations in this book).

Working with Place Value Disks and Strips

As with most manipulatives and visuals, you may want to allow your students a little time to "play" with place value disks and strips before you begin teaching with them. Of course, you'll want to give your students some parameters for using the materials too. In my classroom, my students and I discuss specific procedures, such as getting the disks out, using them respectfully, and putting them away. This means using the disks only when asked and taking their hands off the disks when someone is instructing or explaining her work. In addition, each student is responsible for organizing his disks by place value when a lesson begins and for making sure the disks are put away in bags just as he found them. Putting procedures in place from the beginning will save you instructional time later.

The final stage of allowing students to record abstractly comes only after (1) there has been plenty of practice with manipulatives, (2) there has been subsequent practice with pictorial representations, and (3) students are able to explain what they are doing and why. You may feel tempted to ask students to jump quickly to abstract computation, but in the end this will rarely build student understanding.

I Do, We Do, We Do, We Do, You Do

At each stage in the C-P-A approach, small-group and partner work is important for facilitating language growth in mathematics. Mathematical learning shouldn't happen in isolation. Interaction with other students provides opportunities for children to explain and justify their work, which is a much higher level of thinking than just "getting the answer." Using a guided release of responsibility model works well for many math lessons. The teacher begins by modeling and thinking aloud about a strategy (I do), students then practice with the teacher (we do), students practice with small groups (we do) and partners (we do), and finally students practice independently (you do). I like to call this the "I do, we do, we do, we do, you do" method of instruction.

"I'm Thinking . . ."

Teacher modeling of mathematical language is crucial if we want our students to use mathematical language themselves. Regarding this language component of math instruction, Keene and Zimmermann's research in reading (Keene and Zimmermann, 1997) can be applied. They state that it is necessary for the teacher to share her thinking first. This gives children clear, explicit language for talking about a concept. Mathematical thinking and language promote understanding that math is more than memorization and a set of rules.

As you teach a strategy, make sure you model the strategy multiple times. And as you model, be sure you think aloud at each step so that students know where you are getting your ideas or why you are doing a particular step. At first, this would mean showing your students what to do with manipulatives, and later it would include demonstrating how to record work on paper. Tell kids to listen carefully for the steps you're sharing but also the language you are using. Using phrases like "I'm thinking . . ." or "I'm wondering . . ." or even asking students "What are you thinking?" helps to remind students that you expect them to be thinking critically about what they are doing.

Math Is Not a Spectator Sport

Mathematics is not a spectator sport. Students require daily opportunities to develop mathematical language. After you've modeled a new strategy, encourage your students to talk about what they are doing as they begin practicing the strategy themselves. You'll notice throughout this book that the examples I provide include plenty of questions. I routinely prompt my students to explain their thinking by asking questions like "How did you solve this equation?" or "Why did you solve the equation that way?" I might challenge them by saying, "Can you solve it another way? How do you know your answer is correct?" Or, "Can you justify your method?" My students also know that I expect them to pose questions to me and to each other about the mathematics they are doing, by asking "What would happen if . . . ?" or "Why did we . . . ?"

Your classroom is not going to be a quiet place when your students are using their math talk. That's a good thing! As you walk around coaching your budding mathematicians, stop and listen to their language. The research of Vygotsky (1978) as well as Wegerif and Mercer (1996) argues that students need to engage in "thinking aloud talk." This talk allows the students to explore, clarify, and even question concepts. It also provides opportunities for you to evaluate your students' understanding or misunderstanding of a topic.

Encourage your students to use correct vocabulary when they are explaining their work or asking questions. Here's just one example. When students are talking about expanding numbers while using the left-to-right strategy for addition, you'll want to hear them say something like: "I'm starting with a total of 345. If I expand it [expanded notation], then it becomes 300 + 40 + 5." Or, when adding 345 + 278, you want to hear students say something like: "First, I'm going to add the hundreds. That means 300 + 200." (Not "3 + 2"!) "300 + 200 = 500. Now I need to add the tens. That's 40 + 70. I can make 100 and there will be 10 left. So 40 + 70 is 110." And so on. When students share their thinking, it will lead to improved mastery and an ability to apply the skill to other expressions and situations.

Occasionally, stop the entire class and share some great language or thinking from a group or individual. The mathematical conversations that students have with you and their classmates will provide many occasions for higher-level thinking. This follows Bloom's Taxonomy as well. Students begin by thinking at the lowest knowledge level. As they move from knowledge to comprehension and all the way up to evaluation, they have raised their level of thinking. Coach your students to ask their classmates and themselves why they have done a specific step or used a certain strategy. Get your kids questioning what they are doing: the who, what, when, where, and why of mathematics.

It's the Same!

Along with using correct mathematical language, making connections between strategies is critical! Students need to "see," for example, that the left-to-right strategy and use of place value disks are really the same thing written in different ways. Give students a chance to discover these connections on their own before you explain them to your full class. It's great to hear the "Oohs" and "Aahs" when students see these connections. I am always delighted when I hear a student say, "Aah! Look! These two are the same." For students struggling to see the connections, placing the different strategies or methods side by side with the same equation helps. Ask kids, "What are you noticing about these two methods?" You're hoping they'll say, "I have a connection . . ." or "This reminds me of . . ." or "This is just like the problem that we did yesterday." At this point I often play devil's advocate and say, "What? Are you sure?" Or, "What do you mean? I don't understand." Or, "I don't believe you." They love to prove to me that they are right! Once a student begins to share the connection, make a big deal out of it!

Computation skills build on one another, and students must have opportunities to learn and practice a variety of strategies that build connections and therefore comprehension. You can facilitate this by presenting visuals of different strategies side by side and intentionally pointing out or having students notice what is the same and/or different about the strategies. The goal is for students to grasp how one strategy relates to another. Not only will students relate various strategies for addition, but they will also make connections between addition and multiplication or between multiplication and division.

Place Value Is Fundamental

A concluding thought about teaching computation: As you begin planning for instruction of each computation method, remember to emphasize place value. Place value should not be taught as a separate topic unto itself—it should be a repeating theme throughout math lessons. Without the concept of places and values in our base ten system, computation will be difficult and long-term understanding nearly impossible. Throughout this book, you'll find place value concepts and language woven into the guided conversations for the various strategies presented. Learn to be fluent in "place value talk," and be sure your students are too!

How to Use This Book

PRACTICE! PRACTICE! PRACTICE! The best way to master math strategies is to practice their use. In each chapter in this book—Addition, Subtraction, Multiplication, and Division—I've presented a recommended progression of strategies for teaching that computation skill. (I also encourage you, though, to dip in and try a particular strategy that you feel may help a student who's struggling with a specific concept.) Each section begins with an introduction to the strategy being taught, followed by one or more in-depth examples of the strategy, including a guided conversation that models the type of conversation I would have with my students while teaching the strategy. Following each guided conversation, the Building On sections provide suggestions and examples for extending the learning.

The methods presented in this book encourage the use of instructional materials. You'll want to have the following materials available for use by teachers and students.

- Place value strips
- Place value disks
- Place value chart
- Number-bond cards
- Part-whole cards
- Decimal tiles
- Decimal strips

You can order place value disks and strips from Crystal Springs Books (www. SDE.com/crystalsprings). Another option is to create your own strips by cutting sentence strips into different lengths for ones, tens, hundreds, and so on. You can cut circles of colored paper to use as place value disks, or even use bingo chips of different colors—one color for each of the different place values you need to represent.

For use with the disks, you'll need to create your own place value charts. Make one for yourself to use as a model as well as one for each student in your class. It's easy to do using an 8½ x 11-inch sheet of unlined paper, as shown in the illustration at right. To make a similar chart for use with three-digit numbers, draw vertical lines that divide the paper into thirds.

You may want to make each student a place value chart that has *X*s in the ones column set up like a ten frame, as shown in the following illustration. This will help with tens combinations.

	X X X X X X X X X X

Creating a math word wall will encourage your students to use correct math vocabulary. To get your word wall started, refer to the reproducible vocabulary words on pages 110 through 115. These pages are especially designed to be photo-copied. You can laminate the copies, cut them apart into individual words, and store them in a folder for use as needed.

Building Understanding

I hope you will find that both your mathematical understanding and that of your students rise to a higher level as you begin to use Singapore Math strategies. Celebrate the multiple ways computation problems can be solved, and remember the goal of your mathematics instruction: to build mathematical understanding that students can take out into the world with them. After all, mathematics is not just about right answers. It is about exploring different ways of thinking. It's about teaching your students that if they understand the patterns and relationships, they can solve anything. Now, let's get going! It's time to develop your mighty mathematicians . . . one strategy at a time.

Addition

ADDITION IS A CONCEPT STUDENTS BEGIN TO USE long before they start school. As toddlers, they ask for more cookies, look at more books, or want more crayons. In effect, they have a portion (part) of something and they want more (another part) so that they will end up with a larger total. As teachers, we can use our students' life experiences when we teach the concept of addition. In other words, we can help students to realize that they already know about addition. They have experienced the concept of Part + Part = Total in everyday life, and now we are introducing terminology that will help them apply the concept as they work with mathematical expressions.

Start by emphasizing the part-whole relationship of the numbers. If you know two parts, you can compute the whole (5 + 6 = ___). If you know the whole and one part, you can find the other part (5 + ___ = 11). That's the basic concept behind all addition. Once students understand part-whole thinking, you can introduce strategies that ask them to expand and decompose numbers. That sounds like the distributive property to me! Starting with number bonds and part-whole thinking will help young mathematicians make that important connection to the distributive property. Understanding number bonds and the part-whole concept is fundamental to building number sense in your students. Make sure they've mastered both before you move on to the rest of the computation strategies in this chapter.

Manipulatives and visuals can be a big help in reinforcing an understanding of number bonds and part-whole thinking. You'll want to purchase place value disks so that students can build and decompose numbers. Two visuals you may wish to purchase or create are number-bond cards and part-whole cards. These cards emphasize the many different ways that numbers can be decomposed or bonded. Students can use them to explore the relationships between numbers.

Connecting the addition strategies you teach is crucial so that students understand that these strategies are related and know how to use them together. For example, students need to learn that number bonds and decomposing make addition easier when using left-to-right, place value disks, vertical, and traditional methods. Guide your students to notice what is the same and different about these methods.

As you teach addition, always keep in mind that place value is essential to the concept of regrouping (you may also call this trading or renaming). When you begin, use expressions that involve *no* regrouping (such as 23 + 34); then move on to expressions that require regrouping (45 + 48, for example). Help kids understand the basic structure of the method before adding the trickiness of regrouping. After teaching both without regrouping and with regrouping, be sure to mix it up. You want your students to think afresh about what they are going to do to solve each addition problem they encounter.

Ultimately, mathematics is all about problem-solving. We teach students computation for addition so that they can apply the strategies to math stories (word problems) or, more importantly, to real-life situations in which items are being put together. By following the stages laid out in this chapter, you'll give your students many different strategies or tools for solving addition equations. Encourage your students to continue to use a variety of strategies. One way to do this is by applying several addition strategies daily to math stories. After completing a math story, ask students to share the computation methods they used to find a solution. Celebrate the diversity of strategies and mathematical thinking.

Stages of Addition

Remember the basics: begin with the concrete, move on to pictorial representations, and end with the abstract. You can do this by teaching the concept of addition in the following sequence:

1. Number bonds

2. Decomposing numbers

3. Left-to-right addition

4. Place value disks and charts

5. Vertical addition

6. Traditional addition

Number Bonds

NUMBER BONDS HELP STUDENTS SEE that numbers can be "broken" into pieces to make computation easier. With number bonds, students recognize the relationships between numbers through a written model that shows how the numbers are related. The relationship is visually displayed in one simple bond. This practice is a change from that of traditional math textbooks, which instructed students to memorize all of their number facts as separate entities. Students learn that the numbers 3, 7, and 10 are all connected instead of having to memorize by rote that $3 + 7 = 10$, $7 + 3 = 10$, $10 - 3 = 7$, and $10 - 7 = 3$. Students also learn that there is more than one number bond for 10. It's all about manipulating, or "playing with," numbers!

You'll notice that my example plays with the number 10. That's because understanding the concept of 10 is so critical for place value and computation. Giving students a solid foundation in number bonds from 1 to 10 is a fundamental part of the Singapore Math approach. I strongly suggest that you ensure your students focus on the concept of numbers to 10 before you begin teaching the addition strategies in this chapter. Student understanding of number bonds to 10 lays the foundation for more difficult computation and place value strategies.

INTRODUCING THE STRATEGY

Guided Conversation

Step One: Select a number and write it down.

10

Step One: Today we're going to play with different ways to make the same number. We'll call the families of numbers we make number bonds. What will we call them? Yes, you're right. Number bonds. We're going to start with the number 10. We've talked about how important this number is, so we want to become experts on the number 10. One more time, what are these families of numbers called? Number bonds.

Step Two: Think of a way to break up the number into parts. What are two addends (parts) that can go together to make the total? Show this with counters. Use a straw, pencil, or wooden craft stick as a divider to create a visual split between the groups.

Step Two: I want to think of some different ways to make 10. Hmmm. Let me get out 10 counters. I'm going to split them into two groups. One way I can split these 10 counters is to make a group of 6 counters and a group of 4 counters. I know that 6 and 4 make 10. 6 is a part. 4 is a part. 10 is the total. That means 4, 6, and 10 make a number bond.

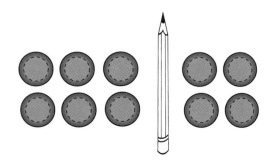

Step Three: Find other ways to decompose (break into parts) the total. What are they?

Step Three: Let me put the 10 counters back together. Can you think of another way to break up the number 10? Who'd like to come up and show us another way? You noticed that 7 and 3 make 10. 7 is a part. 3 is a part. 10 is the total. That's right—thank you! Can anyone find another way? 5 and 5? So 7 + 3 is equivalent to 5 + 5. That's great!

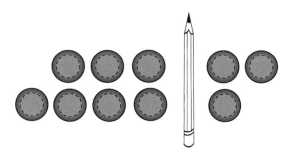

Step Four: Act out a number bond for the chosen number, and then write the number bond.

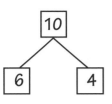

Step Four: Now I want to show you how we can write these number bonds on paper to show the parts and the totals. But first, we are going to act out a number bond. What was one way we said we could make 10? Oh yes, 6 and 4 make 10. Watch me and copy my actions. Put one hand out and say, "One part is 6." Put your other hand out and say, "The other part is 4." Put your hands together over your head and say, "The total is 10." Do it again: "6 [one hand out] and 4 [other hand out] make 10 [hands over head]." Look how I write this. It looks just like our motions!

Guess what? I can do the motions backwards too! Put your hands together over your head. 10 is the total. Move one hand down—one part is 6. Move the other hand down—the other part is 4.

Building On

Understanding number bonds is critical for students to be successful with all the strategies that will follow. Once you've introduced this basic concept, you'll want to give your kids lots of practice. Start with the simplest combinations, and then work up to more difficult ones. Here are some examples you might use.

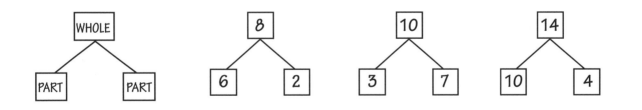

Next, begin with the parts and ask your students to tell you the whole.

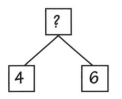

Help your students discover how they can make number bonds for numbers with greater value too. Start with examples showing the whole and the parts. Then give them the whole and ask them to tell you the parts.

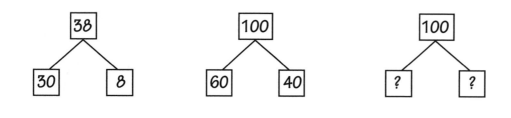

Number bonds can be used with decimals, fractions, measurement, and time too!

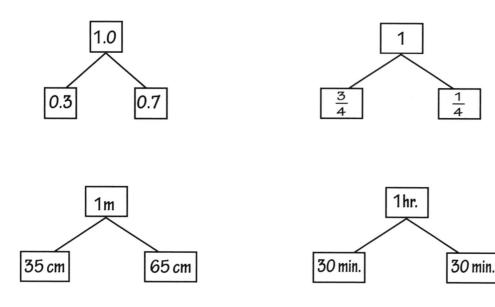

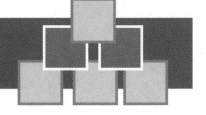

Building On and On

Number bonds can be used with both addition problems that don't need any regrouping and those that do require regrouping. Below are examples that show regrouping of the ones and tens, and even decimals. Keep in mind that using number bonds encourages students to look for friendly numbers like 10, 100, and $1.00. (When addition is the topic in question, friendly numbers are multiples of 10, 100, and so on.) Using these numbers encourages students to exercise their mental math skills. For example, given the expression 38 + 47, a student might notice that 38 is only 2 away from 40. She could decompose 47 into 2 and 45 and add the 2 to the 38 to make 40. That leaves her with 40 + 45, which is much easier to solve mentally than 38 + 47. Notice in the following examples how number bonds have been used to solve problems with greater-value numbers by making groups of 10 or 100. In the final stage of reaching the solution, encourage students to use mental math to find the sum.

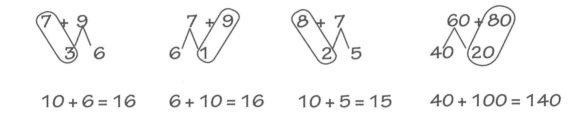

$$10 + 6 = 16 \qquad 6 + 10 = 16 \qquad 10 + 5 = 15 \qquad 40 + 100 = 140$$

$$88 + 23$$
$$2 \wedge 21$$
$$(88 + 2) + 21$$
$$90 \quad + 21$$

Think: $(90 + 20) + 1 = 111$

$$163 + 49$$
$$162 \wedge 1$$
$$162 + (1 + 49)$$
$$162 + \quad 50$$

Think: $(160 + 50) + 2 = 212$

Is there more than one way to employ number bonds to find a solution? Absolutely! That's the beauty of this strategy. One student may see 162 + 50, while another may see 160 + 52. Either way, they are playing with numbers and place value to find the sum.

You can even apply the concept of number bonds to addition expressions that involve time, money, and fractions! The "trick" in these cases is deciding what represents a whole unit. For example, 1 would be the whole for expressions that include fractions, while an hour (60 minutes) would be the whole for expressions involving time. Think about how you could use this same strategy for expressions involving length, weight, and more. I love it when I find one strategy that can be used in so many different ways!

$$\frac{3}{4} + \frac{1}{2}$$

$$\frac{1}{4} \bigwedge \frac{1}{2}$$

$$\frac{1}{4} + \left(\frac{1}{2} + \frac{1}{2}\right)$$

$$\frac{1}{4} + 1 = 1\frac{1}{4}$$

$$\frac{5}{8} + \frac{6}{8}$$

$$\frac{3}{8} \bigwedge \frac{3}{8}$$

$$\left(\frac{5}{8} + \frac{3}{8}\right) + \frac{3}{8}$$

$$1 + \frac{3}{8} = 1\frac{3}{8}$$

$$\$1.35 + \$0.85$$

$$\$1.20 \bigwedge \$0.15$$

$$\$1.20 + (\$0.15 + \$0.85)$$

$$\$1.20 + \$1.00 = \$2.20$$

$$\$0.66 + \$0.75$$

$$\$0.41 \bigwedge \$0.25$$

$$\$0.41 + (\$0.25 + \$0.75)$$

$$\$0.41 + \$1.00 = \$1.41$$

---

45 min. + 45 min.

15 min. $\bigwedge$ 30 min.

(45 min. + 15 min.) + 30 min.

60 min. + 30 min. = 1 hr. 30 min.

1 hr. 35 min. + 50 min.

25 min. $\bigwedge$ 25 min.

(1 hr. 35 min. + 25 min.) + 25 min.

2 hr. + 25 min. = 2 hr. 25 min.

Decomposing Numbers

DECOMPOSING NUMBERS ENCOURAGES STUDENTS to think about place value, and students' awareness of place value will be key to later success with mental math and other methods of computation. Decomposing builds on the concept of number bonds discussed in the preceding section of this chapter, reinforcing the connections among number bonds, place value, and multidigit numbers.

Decomposing numbers is best taught using place value disks and strips, or even base ten blocks if those are the manipulatives you have available. As always, make sure that as students work with manipulatives, draw pictures, and move to abstract computations, they use the correct vocabulary. Keep them talking about ones, tens, hundreds, place value, regrouping, and trading!

You will want to work through many examples with your students. Ask your students: "What do you notice about the disks and the strips?" Try to let kids "discover" what's happening with the ones and tens instead of giving them all of the information. For example, let them discover that 7 tens disks plus 3 ones disks is the same as a 70 place value strip plus a 3 strip. And let it unfold that the principles continue to apply as they begin working with numbers of greater value. The lesson is much more powerful and impactful that way. After working with a few examples, students should begin to make connections and recognize that they're using the strips and disks to build the same number. They're really showing the same thing! If kids aren't noticing anything, you might want to prompt them by mentioning that you're seeing a pattern. Encourage them to be mathematical detectives.

INTRODUCING THE STRATEGY

Step One: Select a number and expand the number using place value disks or base ten blocks. Think: How many ones? How many tens? How many hundreds?

76

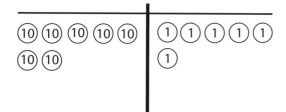

Step Two: Expand the number using place value strips.

Step Three: Put the strips back together again to show what the total is.

Step One: We've been working on building numbers with our place value disks. Do you think we can build the number 76? Sure we can! I see a 7 and a 6 in the number 76, but I know the 7 means 70, or 7 tens. Let me grab 7 tens since the 7 is in the tens place. And let me put those disks in the tens column on the place value chart. Can you help me count by tens as I do that? 10, 20, 30, 40, 50, 60, 70. Now I need 6 ones, because the 6 is in the ones place. Can you help me count those out too? 1, 2, 3, 4, 5, 6. Great! Your chart should look like this.

Step Two: Now that we've built the numbers with disks, we're going to use our place value strips to match. We're going to explore how to expand a number by place value. Let's try the same number: 76. Make sure you have your place value strips ready to go.

How can I make the number 76 with place value strips? I know that I see a 7 and a 6. Is the 7 really worth 7? No, it's worth 70. Why? Because the 7 is in the tens place and 7 tens is the same as 70. What about the 6? Is it worth 6? Yes. Why? Because the 6 is in the ones place. So if we expand 76 into parts by place value, we'll get a 70 and a 6.

Step Three: Let's use our place value strips to prove that 70 and 6 make 76. Slide your 70 and 6 strips back together. What do you get? 70 and 6 make 76. I wonder what would happen if I switched them around. What if I slide 6 and 70 together? Will I get the same thing? Yes, it's still 76.

Step Four: Starting with the total, pull the strips apart again to "prove" the parts.

70   6

Step Four: Now hold up your 76. What two parts will you have when you pull the strips apart? 70 and 6. Are you sure? Why? It's the reverse of what we just did when we put them together.

Building On

Once your students have discovered the basic patterns involved in decomposing numbers, build on that understanding. Repeat the process with three-digit numbers. After you've worked through a couple of three-digit examples, have students explain to you what they are doing and why. Students should be able to justify their thinking.

Repeat the process, this time with four-digit numbers, again making sure students can explain their thinking. Then mix it up! Tell kids that now you're going to see what they have learned about place value and how important it is. Start with a decomposed or expanded number that is not in the correct sequence. Ask students to orally tell you the number or write it down as we traditionally write it. Ask, "Can you explain how you got your answer? How do you know?" Or, "What did you do? Why? Prove it with strips!"

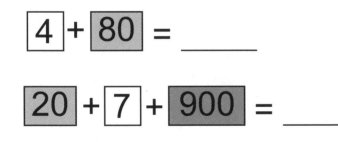

$$4 + 80 = \underline{\hspace{2cm}}$$

$$20 + 7 + 900 = \underline{\hspace{2cm}}$$

This strategy works beautifully for students in higher grades too. You can introduce decimals by using decimal strips and tiles. Following the same process you used for whole numbers, you might ask students to work through examples like these:

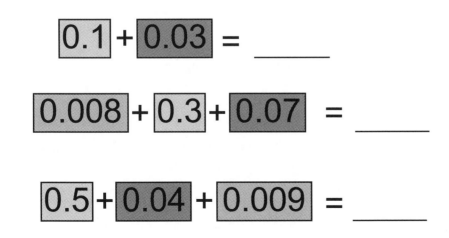

$$0.1 + 0.03 = \underline{\hspace{2cm}}$$

$$0.008 + 0.3 + 0.07 = \underline{\hspace{2cm}}$$

$$0.5 + 0.04 + 0.009 = \underline{\hspace{2cm}}$$

Left-to-Right Addition

AFTER STUDENTS HAVE PRACTICED DECOMPOSING NUMBERS with place value disks and strips, begin left-to-right addition. The goal of this method is to emphasize place value; instead of using the traditional approach of adding the ones and then the tens, this strategy teaches students to work from left to right. The decomposing strategy is an important prerequisite because in order to do left-to-right addition, your students will need to decompose each number.

After you introduce this method and students are feeling confident with it, they'll try finding the sum by using place value in different ways, especially with numbers with greater values. Ask students to share how they found an answer. For example, for the expression 25 + 36, students would begin by expanding the numbers to get (20 + 30) + (5 + 6); that becomes 50 + 11. One student might say 50, count by tens to get 60, and add the final 1 to get 61. Another may figure that 2 tens + 3 tens = 5 tens, which is the same as 50. Then he'll look at 5 + 6 and notice that that is doubles plus one (5 + 5 + 1). Celebrate the different ways of finding the sum. Ask students why they did what they did. Once students have mastered left-to-right addition on paper, challenge them to try to use it mentally without recording all of the steps. Good luck!

You may have noticed by now that many of the addition expressions in the strategies I've explained are written horizontally. This is intentional. Writing addition and subtraction expressions horizontally encourages students to use their mental math skills—in other words, to think by place value. Addition expressions that are written vertically encourage students to follow traditional rules and methods. Go for it! Try horizontal expressions. You'll be pleased with your students' success when they start thinking by place value.

Guided Conversation

Step One: Write the addition expression horizontally.

34 + 45

Step One: We've gotten so good at decomposing numbers by place value. Now it's time for us to use these skills to help with addition. I'm going to write the expression horizontally. That's like lying down. Use your arms to show me what *horizontal* looks like. That's right; your arms should be extended straight out from your sides.

Step Two: Decompose each number by place value. Put the tens together and the ones together. Be sure to write down these parts when learning this strategy!

(30 + 40) + (4 + 5)

Step Two: Now I need to decompose by place value. Remember that means I'm going to pull the numbers apart into tens and ones. Let's start with 34. How can we break apart the number 34? Everyone show me with your place value strips. That's right. 30 and 4. Now show me 45. Yes, it can be pulled apart into 40 and 5. Now I'm going to put the tens together with parentheses and then do the same thing with the ones. Team up with a neighbor to show me how the decomposed numbers can be put together. Wonderful!

Step Three: Add the tens and ones and record each total separately.

(30 + 40) + (4 + 5)
 70 + 9

Step Three: We just finished pulling apart the numbers, and now we need to add them together. Start with the tens. Which numbers will we add first? That's right; 30 + 40. What is 30 + 40? 70. How do you know? Because 3 tens + 4 tens = 7 tens, and that's the same as 70. Record the 70 underneath the 30 + 40. What's left? The ones. Let's add the ones. 4 + 5 = 9. Record the 9 under the 4 + 5.

Step Four: Find the sum of the subtotals.

$$(30 + 40) + (4 + 5)$$
$$70 \quad + \quad 9 \quad = 79$$

Step Four: We've added the tens and the ones separately, and now we need to put them all back together. What do we have? We have a 70 and a 9. What is the sum of 70 and 9? 79. Prove it with your place value strips. Which strips do you need? The 70 and the 9? That's right! We pulled the numbers apart before, didn't we? So now we have to put everything back together. What do we get? 79. Good work!

Building On

Give kids plenty of time to practice using the left-to-right strategy and to solidify their place value work before moving on. Students will naturally differentiate their computation when using this strategy. Some will choose to add the ones first; others will start with the tens. When you introduce numbers with greater values, kids will find all sorts of creative ways to compute the final sum. Ask your students to explain to their classmates the process they used. This will encourage others to try new approaches they may not have thought of before. Celebrate their unique ways of thinking. Eventually, you may have some students who can complete the entire process mentally. Wow!

As students gain confidence, introduce problems that require regrouping, as well as problems with three- and four-digit numbers.

$$67 + 28$$
$$(60 + 20) + (7 + 8)$$
$$80 \quad + \quad 15$$
$$90 + 5 = 95$$
$$67 + 28 = 95$$

$$145 + 348$$
$$(100 + 300) + (40 + 40) + (5 + 8)$$
$$(400 \quad + \quad 80) \quad + \quad 13$$
$$480 \quad + \quad 13$$
$$490 + 3 = 493$$
$$145 + 348 = 493$$

2,247 + 6,172
(2,000 + 6,000) + (200 + 100) + (40 + 70) + (7 + 2)
   8,000   +   (300   +   110) + 9
   8,000 +   (410   +   9)
   8,000 +   419
8,000 + 419 = 8,419
2,247 + 6,172 = 8,419

Building On and On

The left-to-right strategy can be used to solve equations involving decimals and money too.

3.6 + 9.8
(3 + 9) + (0.6 + 0.8)
  12 +   1.4
12 + 1.4 = 13.4
3.6 + 9.8 = 13.4

$4.68 + $2.72
($4.00 + $2.00) + ($0.60 + $0.70) + ($0.08 + $0.02)
   $6.00   + ($1.30   + $0.10)
    $6.00   + $1.40
$6.00 + $1.40 = $7.40
$4.68 + $2.72 = $7.40

Place Value Disks and Charts

EARLIER IN THIS CHAPTER, we used place value disks and strips to build students' understanding of place value by decomposing numbers. We then applied this learning to the left–to-right strategy for addition. Now we are heading toward the traditional algorithm, with the goal of ensuring that students understand the mathematics behind this algorithm, rather than simply memorizing a set of rules. To accomplish this, we are going to use place value disks once again. In many ways, this simulates the traditional method for addition that most of our parents learned when they were children.

Through manipulating the disks, though, kids will learn when and why they should use regrouping. Be sure you begin with problems involving one- and two-digit numbers only, because students need to understand the concept of regrouping (renaming or trading) before working with numbers with greater values. For some students, regrouping is a hard concept to master. They often want to grab all of the ones disks they have and trade them in for a tens disk. Make sure they trade in only 10 ones and leave the extras.

In my class, my students and I always chant, "Start in the ones." I ask my students: "What does it say (referring to the ones place)? Do I have enough ones to make a trade?" After deciding whether there are enough ones to make a trade and completing the action with the disks, we move on to the tens. Then we chant, "Go to the tens." I ask: "What does it say (referring to the tens place)? Do I have enough tens to make a trade?" Again we talk through what we are doing with the disks, decide whether to make a trade, and discuss why or why not.

With this strategy, begin by asking students to manipulate the disks only. Help students to see when they do or do not need to regroup. Ask why. Can they "prove it"? As they work with the disks, you can model how to record on paper (abstract) what is happening with the disks. Once students are comfortable working with disks, have them progress to making drawings of disks and simultaneously recording their work. This will build the connection from the concrete to the pictorial to the abstract.

Step One: Review what the columns on a place value chart represent. Then write an addition expression, using one- or two-digit numbers.

$$\begin{array}{r} 45 \\ + 26 \\ \hline \end{array}$$

Step Two: Using place value disks and a place value chart, use disks to build the two numbers, placing ones and tens in the correct areas on the place value chart.

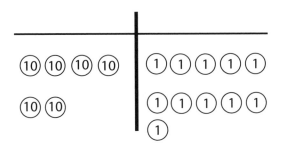

Guided Conversation

Step One: We've learned several methods for solving addition equations, and today we're going to use place value disks to really show us what is happening to the ones, tens, and even hundreds when we add. We're also going to use our place value chart to work on. Why do you suppose there are three columns on the chart? You're right. There is one column each for the hundreds, the tens, and the ones.

Now, to get started, I'm going to write an addition expression. Up until today, we've been writing most equations horizontally, or lying down. Today, though, we're going to write our equation vertically, or standing up. Everybody use your arms to show me *horizontal*. That's right. Now show me *vertical*. Good, you've put your arms straight up over your head.

Step Two: Let's look at the first number, 45. We need to build the number 45 with our place value disks. How will we do that? Yes, we need to get out 4 tens and 5 ones. Remember to set them up in a row as you would in a ten frame. A ten frame uses two rows with five spaces in each row. That way it's easier to count what's in each row. If I look at your place value chart, I should see 4 tens disks lined up in a row in the tens column and 5 ones disks lined up in a row in the ones column to represent the number 45.

Now we're going to build the second number, 26, underneath. How will we do that? Yes, get out 2 tens disks and 6 ones. How will we set up these disks? We'll put the 2 tens in the tens column. Then we'll put the 6 ones disks in a row of five and a row of one in the ones column. We've now built our two addends. If we are adding, what are we going to do? Put them together.

Step Three: Add the disks together by column, starting with the ones. Add the ones together. Discuss whether renaming (regrouping, trading) is necessary. If no, move on. If yes, make the trade with your disks. Be sure to discuss what you are doing and why.

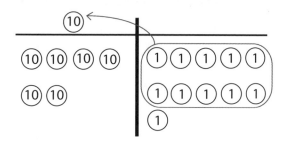

Step Four: Next, move to the tens place. Add the tens together. Ask whether renaming is necessary. Discuss why or why not. If no, move on. If yes, use your disks to make the trade.

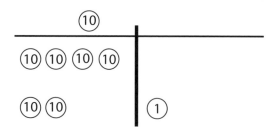

Step Five: Count how many tens disks and how many ones disks you now have. Record the sum of the two numbers.

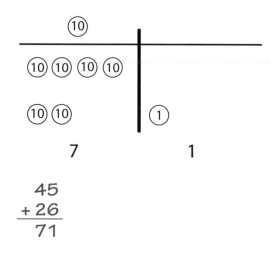

$$\begin{array}{r} 45 \\ +\,26 \\ \hline 71 \end{array}$$

Step Three: Now we're going to add the disks together starting in the ones. If we look in the ones, it shows 5 + 6. Do I have enough ones to make a trade? Yes. How do you know? How many do you need before I can trade? That's right; I need 10. In this case, I have 11 ones. I have enough to trade. So, what do we do? Grab 10 ones and trade them in for a ten. Watch. Look where I put the ten. Why did I put it in that place? You got it. It's a tens disk, so I need to put it in the tens place. Do you have any ones left? Yes. Why? Because we started with 11 and we traded in only 10.

Step Four: Next, we'll move to the tens since they have the next-greater value. What does it say in the tens place? It says 4 tens + 2 tens + 1 ten (from the trade). Do I have enough tens to make a trade? No. Why not? That's right. We need at least 10 tens to make a hundred. We have only 7.

Step Five: Let's find our sum by counting the tens and the ones. How many tens do we have now? There are 7 tens. Let's record that. How many ones? There is 1 one. Let's write that down. What's our total? Now can you remind me: what is the value of 1 ones disk? 1! That's right! How about 7 tens? It's not just a 7, is it? The 7 is a 70! Why? It's a 70 because the 7 is in the tens place. So we have 7 tens, which equals 70, and 1 one, which equals 1. What is 70 + 1? 71.

After your students have practiced using place value disks and can explain how they found answers, then they are ready to begin drawing (pictorial stage) while they work with the manipulatives. Be sure students work one step at a time. First, they will build the two addends with disks and then draw them. Next, they will add the ones, make a trade as necessary, and record that step on their drawing. They will repeat this sequence in the tens place, and so on, until the final step of recording the sum. Remind your students that they should continue to ask themselves questions such as "Do I have enough ones to trade?" even when you are not there to prompt them.

Listen carefully for your students' conversations as they use drawings to solve the addition problems. If you hear students using phrases like "Cross out this and put this," you need to intervene. It's a cue that they are working from the basis of rote memorization of steps and not conceptual understanding. When kids can explain the what, when, and why of the process, then you'll know they understand it. Then they're ready for the traditional abstract algorithm.

$$\begin{array}{r} 55 \\ + 27 \\ \hline 82 \end{array}$$

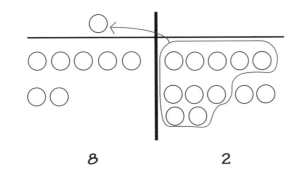

8   2

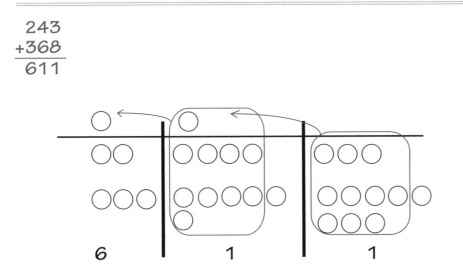

$$\begin{array}{r} 243 \\ + 368 \\ \hline 611 \end{array}$$

6   1   1

Vertical Addition

THIS METHOD IS VERY SIMILAR TO LEFT-TO-RIGHT ADDITION except that—you guessed it—it's vertical instead of horizontal. It is also an easier strategy than left-to-right to use for three- and four-digit numbers. The goal is once again to pay attention to place value. One of the beauties of this strategy is that students can start with any place value and work in any order, finding the partial sums of the ones, tens, and/or hundreds before finding the final sum. It is generally simpler, however, to begin with the largest value so that students can line up the numbers more easily.

Students who are struggling may want or need to use place value disks or strips as they learn this strategy. I find that the place value strips are easier to use. Coach these students to expand the numbers and use the strips to help themselves pay attention to the place value of each digit.

Encourage students to continue to make connections between the various strategies they've learned. After teaching this strategy, you may even want to work with students to write up an example using left-to-right addition, place value disks and charts, and vertical addition on a display and notice all of the similarities.

INTRODUCING THE STRATEGY

Step One: Write the addition expression vertically.

```
   16
 + 29
```

Step Two: Add the tens and record the tens total below the line.

```
   16
 + 29
 -----
   30
```

Step Three: Add the ones and record the ones total below the tens total.

```
   16
 + 29
 -----
   30
   15
```

Step Four: Find the sum.

```
   16
 + 29
 -----
   30
 + 15
 -----
   45
```

Guided Conversation

Step One: Wait until you see the strategy I'm going to show you today. I am hoping you will make a connection to another strategy we have used. Be ready to share what your connection is and why you think this method is connected to the one you're thinking about.

I'll start by writing an addition problem vertically. What does it say? 16 + 29 equals what?

Step Two: First, I'm going to add the tens. Let's see. There's 1 ten, which equals 10, and 2 tens, which equals 20. The total of 10 and 20 is 30. I'm going to write 30 underneath the line. I know how to add 10 + 20 mentally, but if I didn't, I could write it to the side. I did not write 1 + 2 = 3. Why not? That's right; because the 1 and the 2 are in the tens place.

Step Three: Next, I'm going to add the ones. I see 6 and 9. So we need to add 6 + 9. What does it equal? Yes, 15. How did you figure out 6 + 9? You made a ten so that the problem turned into 5 + 10? Brilliant. Who figured it out a different way? Oh! You knew that 6 + 10 was 16, so 6 + 9 would be 1 less, which is 15. Amazing. Now I'm going to write 15 below the 30.

Step Four: Last, I need to find the sum of 30 and 15. It's 45. I figured that out mentally. I knew 30 + 10 = 40, and then I just added 5 more to get 45.

Building On

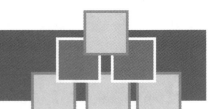

Vertical addition is a great way to stretch place value concepts to many levels, including addition of three-digit and four-digit numbers. After using vertical addition with two-digit numbers, challenge kids to try to use it with three-digit numbers. Students who truly understand place value concepts will be able to make this application with no problem. Ask your kids how they knew what to do. You'll realize they're on track when they reply with something like, "It's just the same thing with bigger numbers." I love it when that happens!

```
    75              459            2,271
  + 46            + 927          + 3,499
  ─────          ──────          ───────
   110            1,300            5,000
  + 11               70              600
  ─────               16              160
   121            +  ──            +   10
                   1,386          ───────
                                    5,770
```

Building On and On

Vertical addition works for equations involving decimals and money too.

```
    9.5            $ 3.79
  + 8.7          + $ 6.28
  ─────          ────────
   17.0            $ 9.00
  + 1.2            $ 0.90
  ─────          + $ 0.17
   18.2          ────────
                  $ 10.07
```

Traditional Addition

THE FINAL ADDITION STRATEGY that students should learn is the traditional method. When you teach this method, make sure you connect it to other methods. In fact, I suggest creating a poster with the students on how to solve an equation (such as 75 + 94 or 317 + 229) using left-to-right addition, place value disks and charts, and vertical addition so they can see the relationships among the strategies. Help students to understand that the traditional method is a shortcut way of showing ones, tens, and hundreds with and without regrouping.

When you begin teaching the traditional method, have your students use disks so they can manipulate and then record, manipulate and then record, manipulate and then record. This helps students understand why they are writing each numeral in each specific location. Be careful about language too: avoid using rote phrases such as "Put a 4 down here" or "Put a 1 up there." You should be using appropriate place value language and so should your kids.

After working on several addition expressions for which students use disks as well as recording what's happening, you may want to put away the disks. A great bridge from the concrete to the abstract is to have your kids pretend or act out grabbing the ones and making the trade for a ten. Remind them that they are doing exactly the same thing they did with the disks, only this time they're writing it down in numbers instead of moving disks.

INTRODUCING THE STRATEGY

Step One: Write the addition expression vertically.

$$\begin{array}{r} 48 \\ +\,97 \\ \hline \end{array}$$

Step Two: Add the ones. Do you need to trade? If no, write down the sum below the ones column. If yes, record one trade (or more if necessary) in the tens and write the remaining number of ones below the ones column.

$$\begin{array}{r} {\scriptstyle 1} \\ 48 \\ +\,97 \\ \hline 5 \end{array}$$

Step Three: Add the tens. Do you need to trade? If no, write down the sum below the tens column. If yes, record one trade (or more if there's more than one ten) in the hundreds and write the remaining number of tens below the tens column.

$$\begin{array}{r} {\scriptstyle 1\,1} \\ 48 \\ +\,97 \\ \hline 45 \end{array}$$

Guided Conversation

Step One: Now it's time to learn the traditional addition method that is probably the one my mom and dad learned when they were in school. To start with, we'll write the expression 48 + 97 vertically.

Step Two: Start in the ones. What does it say? 8 + 7, and we know that equals 15. Do we have enough to make a ten? Yes, we have 15. How many trades? One. What do we trade for? A ten. Why? Because 10 ones equal 1 ten. Where do we record this trade? In the tens. How many ones are left? 5. Where do we record the 5? In the ones.

Step Three: Next, go to the tens. What does it say? 4 tens + 9 tens + 1 ten. Do we have enough to make a hundred? Yes, we have 14 tens. How many trades? One. What are we trading for? A hundred. Why? Because 10 tens equals 1 hundred. Where do we record this trade? In the hundreds. How many tens are left? 4. Where do we record the 4? In the tens.

Step Four: Add the hundreds. Do you need to trade? If no, write down the sum below the hundreds column. If yes, record the trade in the thousands column and write the remaining number of hundreds below the hundreds column.

```
  1 1
   48
 + 97
 ─────
  145
```

Step Four: Now we go to the hundreds. What does it say? 0 hundreds + 0 hundreds + 1 hundred. Do we have enough to make a trade? No. Why? We need 10 and we have only 1. How many hundreds do we have? 1. Where do we write this? In the hundreds place. What's our total? Our total is 145.

Subtraction

SUBTRACTION. THE WORD STRIKES FEAR INTO THE HEARTS of many students because subtraction problems look and sound more difficult than addition. Subtraction with regrouping or involving a number with a middle zero is even scarier. "Look at the ones. Do we need to borrow? Cross out the tens, take a ten, and put that ten together with the ones." Or, "Should we borrow? Cross out the number that you are borrowing from, and put a 1 next to the number of ones." This is the kind of confusing language heard in some elementary classrooms.

What can we do about it? We can help our students understand the concept of subtraction first, rather than bombarding them with abstract rules. Subtraction is about taking items away from a set total. Our students can begin to build an understanding of subtraction by sharing stories in which they have subtracted. These stories could come from reading children's literature or telling their own life stories, such as when a student had 10 pencils and gave 3 to his friend or when a student chose chicken nuggets for lunch, put a total of 6 nuggets on her tray, and then ate 5 of them. How many were left? This is a variation on part-whole thinking. Now the students are beginning with the whole and one part and figuring out the other part. After students understand the meaning of subtraction through stories, you can introduce manipulatives that students will use to build models of subtraction. Next, they will shift to drawing pictures rather than using manipulatives. The last step is to work abstractly, writing subtraction equations and using no manipulatives or drawings. Build comprehension first!

Stages of Subtraction

Subtraction, like addition, begins with a focus on place value and the basic subtraction concept of taking numbers apart before progressing to number bonds. The subtraction concept is solidified using place value disks and charts prior to the traditional algorithm for multidigit subtraction. The following is a recommended sequence for teaching the concept of subtraction, starting with the concrete and ending with the abstract:

1. Number bonds

2. Place value disks and charts

3. Traditional subtraction

Number Bonds

INTRODUCING SUBTRACTION THROUGH NUMBER BONDS is a natural choice because number bonds help students make the connection between addition and subtraction. After all, the first basic step of demonstrating how to break apart a number into parts is very similar to the way your students first learned about addition (see page 13). The emphasis on the part-whole relationship of the numbers remains the same. When they learned about addition, students started with the two parts and then computed the whole or total. Now, with subtraction, students will begin instead with the whole and one part. Their task will be to find the other part. The end result of the two processes is the same: a complete number bond.

It may appear that reintroducing number bonds to teach the concept of subtraction is unnecessary repetition, but don't underestimate its importance. Part-whole thinking (which number bonds practice reinforces) is critical to students' future ability to decompose numbers for use with subtraction strategies involving three-digit numbers and beyond. Part-whole thinking is also involved in many mental math methods.

INTRODUCING THE STRATEGY

Step One: Select a number less than 18 and write it down.

11

Step Two: Think of different ways to break the number into parts. What are two parts that will go together to make the sum? If I know the total and one part, how can I find the other part?

Total = 11
Part = 3
Other part = 8

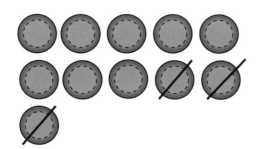

Guided Conversation

Step One: When we were studying addition, remember how we started by playing with different ways to make the same sum? We found out that we can make 10 by adding 3 and 7, or 6 and 4, or 2 and 8. Today we are going to take what we know about addition and connect it to subtraction. Let's start by playing with the number 11.

Step Two: With subtraction, we start out a little differently: we know the total and one part, and we have to figure out the other part. Think of different ways to break the number 11 into parts. Get out 11 counters and show me one way. You broke it up into 3 and 8? What was the total number of counters you started with? 11. How many counters did you "take away"? 3. How many were left? 8. Oh! So you mean 3 and 8 make 11. If we know the total is 11 and one part is 3, then the other part is 8. You've got the idea. Now let's put our counters back together for a total of 11.

Step Three: Find other ways to break up the number into two parts. Is there only one way to break apart 11? What are some other ways? How does knowing the two parts help with subtraction?

Total = 11
Part = 5
Other part = 6

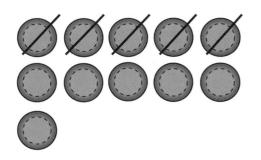

Step Four: Write down the number bonds and notice patterns. Work systematically so that you don't miss any of the combinations.

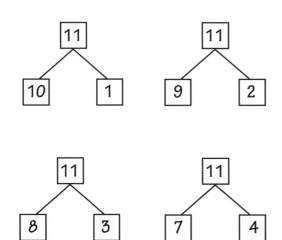

Step Three: What's another way you can break apart 11? Show me what you did. Great! You found that we could also break 11 into 5 and 6. When we were practicing addition, we said that 5 and 6 are parts, and the total is 11. With subtraction, we start by saying the total is 11 and one of the parts is 5. If you didn't know 6 was the other part, how could you figure it out? You're right. You could use counters to show the other part. Another way you could do it is to turn it into an addition problem: 5 + what = 11? If I know 5 + 5 = 10, then I know I need just 1 more, which is 6. Can anyone find another way to break apart 11? Good job! If 11 is the total and one of the parts is 7, then the other part is 4. I wonder how many combinations we can find.

Step Four: We've used our counters to help us figure out how to break apart totals into parts. Now we are going to act it out. What was one way we said we could break apart 11? That's correct. 5 and 6, because 5 and 6 make 11. Now watch and copy my actions. We're going to begin with our hands together over our heads to show the total. Put your hands over your head and say, "The total is 11." Now bring one hand down and say, "One part is 5." Bring your other hand down and say, "The other part is 6." Let's do it again. If 11 is the total (hands over head), then 5 and 6 are parts (one hand out for each number). Now I want to show you how we'd write this down as a number bond. Look how I write this. Next, let's write down all of the number bonds for 11, starting with 10 and 1.

Building On

Number bonds can be applied to numbers with lesser values and to numbers with greater values.

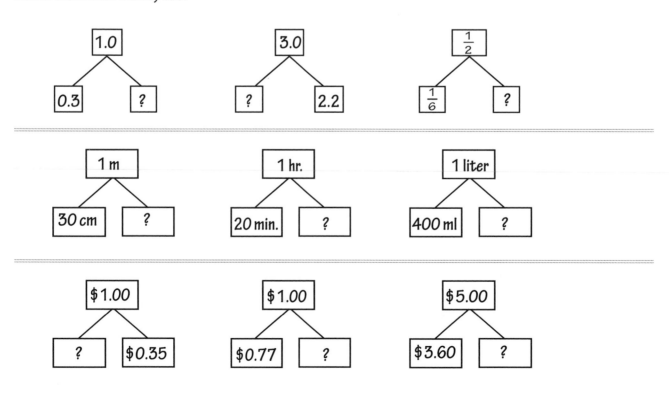

Building On and On

You can also practice number bonds for subtraction of decimals and many units of measurement. Number bonds work with money too!

Place Value Disks and Charts

WORKING WITH PLACE VALUE DISKS AND CHARTS puts the emphasis on the actions needed to solve subtraction problems. Let's look at just one example: 195 – 38. Students will begin by building the whole or total, 195, with their disks. Next, they'll try to subtract 8 ones. They'll quickly notice that this is not possible with the disks they have laid out. They'll need to figure out what action to take to obtain more ones: trading in a tens disk for 10 ones disks. This will give them a total of 15 ones disks, and from there, they can subtract 8 ones and finish solving the problems. (Note that while this example involves a three-digit number, it's best to begin with numbers of lesser value before you introduce problems with higher-value numbers.)

The physical manipulation of trading in a tens disk for 10 ones helps to build understanding that your students can carry through to the traditional subtraction strategy (the abstract phase). In traditional subtraction, students would solve the example of 195 – 38 by crossing out the 9 in 195, replacing it with an 8, crossing out the 5 and replacing it with 15, and then subtracting 8 from 15. With a strong foundation in place value rather than a reliance on rote memorization of rules, they'll understand where the 15 came from and why.

I'll note here how important it is to be careful about our use of precise language when teaching subtraction. For example, I do not advise telling kids that something such as 5 – 8 is "not possible." We all know you can subtract 8 from 5: the answer is negative 3. While you won't want to launch into a lesson on negative numbers in the midst of teaching basic subtraction, you could briefly bring out a thermometer and show kids how the temperature can go down into the negative range. Explain to kids that they do not need to worry about negative numbers now; you just want them to know that they exist. They'll learn about negative numbers when they're in fourth grade!

INTRODUCING THE STRATEGY
Subtracting one-digit from two-digit numbers

Step One: Write a subtraction expression showing a one-digit number being subtracted from a two-digit number.

$$93 - 7$$

Step One: We're going to begin our work with two-digit subtraction today. We've done lots of subtraction facts, but now we're moving to larger numbers, or numbers of greater value. Be ready to use those number bonds we've been practicing. I'm going to write a subtraction expression on the board. What does it say? Yes, 93 – 7. What does 93 – 7 mean? That's right; it means we are going to begin with a total of 93 items and subtract, or take away, 7 of them. How could we visualize what we'll be doing? I'm going to pretend that I have 93 chocolate chips and I'm going to eat 7 of them. What's going to happen to my chocolate chips? You've got it! They're going to go away, so I'll end up with fewer than I started with.

Step Two: Using place value disks and a chart, build the "whole," placing ones and tens in the correct columns. Explain why you do not need to build the second number (because you're taking something apart).

Step Two: We're going to use our place value chart and disks to help us. We need to look at the total, which is 93, and build it with disks. Remind me how to build 93. Yes, we need to get out 9 tens disks and 3 ones disks and place them on our charts. Did you remember to set them in rows of five like a ten frame? If I look at your chart, I should see 9 tens disks, placed in 2 rows. One row will have 5 and the other row 4. I will also see 3 ones disks placed in 1 row on the ones side of your place value chart. Since this is a subtraction expression, what am I going to do? That's right; I'm going to take away some disks. So, do I need to build the second number in the expression, the 7? No, because I am taking 7 away from 93, not adding 7 to it.

Step Three: Look at the ones. Do you have enough ones to subtract? If yes, go ahead and subtract. If no, trade a ten for 10 ones and then subtract.

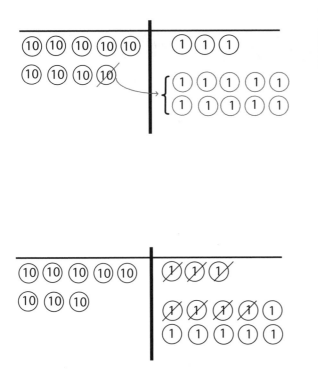

Step Four: Go to the tens. Do you have enough tens to subtract? If yes, go ahead and subtract. If no, trade a hundred for 10 tens and then subtract.

$93 - 7 = 86$

Step Three: Now that we've built the 93, we need to take away 7. Will we be taking away ones, tens, or hundreds? Ones. Why? Because we're taking away 7, not 70 or 700. Place value sure is important! Let's start by looking at the ones. How many do we have? 3. How many do we need to take away? 7. Is that possible with the disks we have? No, we have only 3 ones disks. That means we need to get more ones disks. It's the opposite of addition. Sometimes when we added we ended up with too many ones disks, and we traded them in for a ten. Now we're going to trade in a ten to get more ones disks. So, what should we do? Grab 1 tens disk and trade it in for 10 ones disks. Watch. Notice how I grabbed 1 tens disk and traded it for 10 ones disks. Then I set them up to look like a ten frame. Now you make your trade just like I did.

How many ones did you start out with? 3. How many do you have now? 13. How many tens do you have now? 8. Are you now able to easily subtract 7 ones from 13 ones? Go ahead and subtract the 7 ones. How many ones do you have left? 6. Great!

Step Four: We're through with the ones, so we'll move on to the tens. What do we have in the tens? We have 8 tens. How many do we need to subtract? Zero. Are we able to do that easily? Yes! What is 8 tens minus 0 tens? That's right; 8 tens. That means we have 8 tens and 6 ones left. What is 8 tens and 6 ones? 86.

INTRODUCING THE STRATEGY
Subtracting two-digit from three-digit numbers

Step One: Write a subtraction expression.

234 – 88

Step Two: Using place value disks and a chart, build the "whole," placing ones, tens, and hundreds in the correct columns.

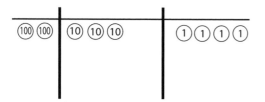

Step Three: Look at the ones. Do you have enough ones to subtract easily (without negative numbers)? If yes, go ahead and subtract. If no, trade a ten for 10 ones and then subtract.

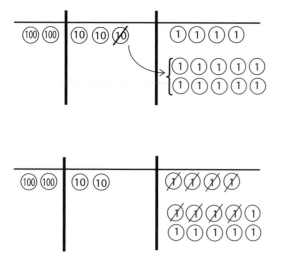

Step One: Wow! We're really getting good at subtraction using these disks. Now I'm going to challenge you to solve 234 – 88. Can you do it? Of course you can. Let's think about what we learned in our earlier practice and see how it can help us here.

Step Two: First, we need to build which number? 234. Should we build 88? No. Why not? Because we are subtracting 88, which means we'll be taking it away, not adding. Excellent. Now how do we build 234? We need 2 hundreds, 3 tens, and 4 ones. Does it matter where I put these disks? Yes, I need to place them correctly on the place value chart. Show me what this should look like.

Step Three: Now we're ready to subtract. The expression says 234 – 88, so we need to subtract 88. Think back to and visualize what we did first when we were solving two-digit problems. Got it? Tell your neighbor what you think we're going to do first. That's right; we're going to subtract the ones. How many ones do we have right now? 4. How many do we need to subtract? 8. Can we do that easily? No, because there are only 4 disks and we need 8. That means we need to get more ones disks. Where do we get more ones from? The tens. What should we do? That's right; we should take a ten and trade it for 10 ones. How many ones did you start with? 4. How many do you have now? 14. Are you able to take away 8 ones now? Yes. How many do you have left? 6.

Step Four: Go to the tens. Do you have enough tens to subtract? If yes, go ahead and subtract. If no, trade a hundred for 10 tens and then subtract. Continue as necessary for other place values.

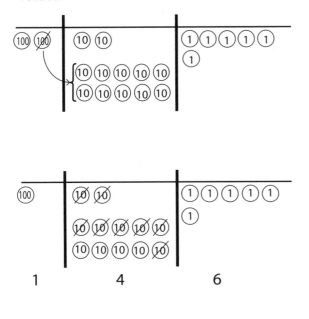

234 – 88 = 146

Step Four: Now that we've finished the ones, what should we do? Yes, we should go to the tens. What does it say in the tens? We have 2 tens and we need to subtract 8 tens. Are we able to do that easily? No. Oh no! We haven't done this before. What do you suppose we are going to do? What did we do when we didn't have enough ones? We went to the tens to get more ones. Where do you think we can get more tens from? That's right; the hundreds. How many tens does it take to make a hundred? 10. Good! We can't subtract 8 tens from 2 tens easily, so we are going to take a hundred and trade it in for 10 tens. How many total tens do you have now? 12. Can you easily take away 8 tens now? Yes. How much is 12 tens minus 8 tens? 4 tens. Check and see. Do you have 4 tens left after you subtract?

What about the hundreds? We don't have any hundreds to subtract, so we can just leave the 1 hundred there. So, what do we have left altogether? We have 1 hundred, 4 tens, and 6 ones, which equal 146.

Building On

Place value disks and charts are excellent tools for building conceptual understanding of subtraction with multidigit numbers, and with decimals too. With practice, students will comprehend that the process remains the same no matter how sizable the numbers become. I actually like to throw in a four-digit problem when my students have been working on three-digit subtraction and ask them if they can do it. If they have been noticing the pattern of what's happening and truly understand the process, they will solve it without hesitation. The same goes for decimals, as shown in this example.

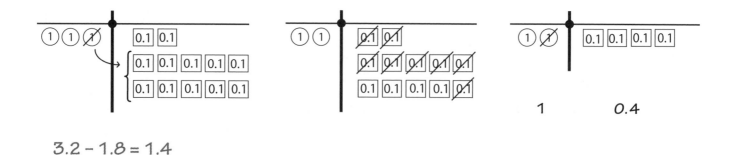

3.2 – 1.8 = 1.4

Building On and On

Once your students are able to manipulate disks and explain what's happening as they work to solve subtraction problems, you can challenge them to move from handling the disks to drawing pictures of the disks. You can begin to model the abstract algorithm as they create their drawings. Advanced students may be ready to start recording what's happening in an abstract algorithm too, working one step at a time. Keep listening too. Are your students using their "math talk" to explain what they are doing and why?

55 - 18

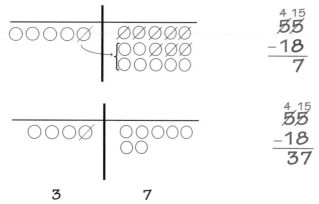

417 – 136

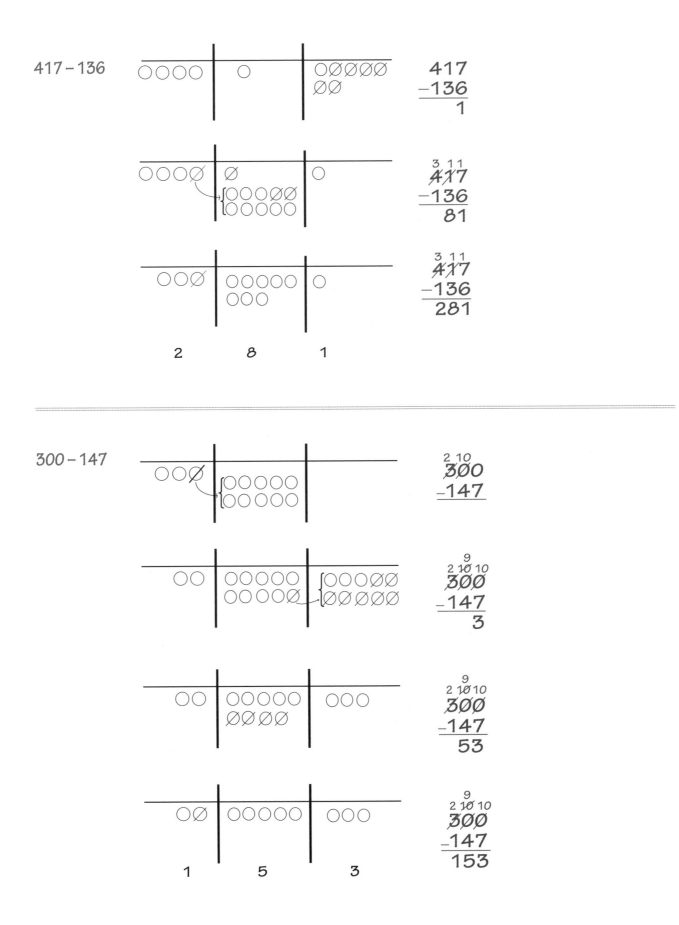

417
−136
    1

  3 11
4̷1̷7
−136
  81

  3 11
4̷1̷7
−136
 281

2   8   1

---

300 – 147

  2 10
3̷0̷0
−147

     9
  2 1̷0 10
3̷0̷0
−147
    3

     9
  2 1̷0 10
3̷0̷0
−147
   53

     9
  2 1̷0 10
3̷0̷0
−147
  153

1   5   3

Traditional Subtraction

MAKING THE MOVE FROM CONCRETE TO PICTORIAL and now to abstract is a big step. It should be done only when students have proven their understanding of subtraction and trading (regrouping) with both manipulatives and pictures.

After your students are skilled at using place value disks and have practiced drawing disks to solve subtraction problems, then it's time for traditional abstract subtraction. Be sure to connect what's happening in the ones, tens, and hundreds places on paper to the actions they took when they were using disks. I remind my kids to visualize what they did with the disks. I encourage them to pretend they are moving the disks if that helps them to visualize the next step in the process.

Introduce the traditional algorithm beginning with subtraction problems that require no trading or regrouping. After plenty of practice, you can progress to problems that do require regrouping. Make sure students talk about what they are doing and why along the way.

I never tire of emphasizing the importance of coaching your kids to use correct language to explain what they are writing and thinking as they solve a traditional subtraction problem. They are not just "crossing out a 2 and putting a 12" or, as many folks do, putting a little 1 in front of the 2 to make it look like a 12. They are making a trade and there's a reason the 2 is becoming a 12. Make sure your students can explain when they should and should not trade as well as why.

As your students work through the traditional algorithm, don't hesitate to stop those who are struggling and ask them to work again with pictures or even manipulatives if necessary. Sometimes I'll draw place value disks to match what a problem says just to help them to visualize what's going on. Often, this little visual reminder is just what kids need.

As you work with single trades in the ones and then in the tens (hundreds too) and your kids gain confidence, try mixing up the problems so that some require trades and others do not. Also make sure that some require trades only in the ones and others only in the tens. This variety encourages students to think about what they're doing and not just trade, trade, trade on every problem in the same way.

INTRODUCING THE STRATEGY

Step One: Write the subtraction problem vertically.

$$
\begin{array}{r}
534 \\
-275 \\
\hline
\end{array}
$$

Step Two: Start in the ones. What does it say? Do you need to trade? If no, write down the difference (the answer to a subtraction problem) below the ones column. If yes, record taking a ten and trading it for 10 ones. Write down the remaining number of tens and how many ones you now have. After trading, subtract the ones and record the difference in the ones column.

$$
\begin{array}{r}
{\scriptstyle 2\ 14} \\
5\cancel{3}\cancel{4} \\
-275 \\
\hline
9 \\
\end{array}
$$

Guided Conversation

Step One: Are you ready for a challenge? We're going to work on subtraction today without using any disks or pictures. Do you think you can do it? Of course you can. We're going to follow the same process that we did with disks and drawings, but we're going to take a shortcut. First, let's write down the subtraction problem 534 – 275 vertically. What does that mean? It means you will write the problem with 534 on top and 275 underneath it.

Step Two: Think back to when we were using place value disks and drawing pictures. What did we do first? That's correct; we started in the ones. What does it say in the ones? It says 4 minus 5. Are we able to do that easily? No. So what should we do? We take a ten and trade it for 10 ones. Pretend you are doing that with disks. Now how do we record that? We cross out the 3 tens and write a 2 above it because we took a ten and now there are only 2 tens left. Next, we cross out the 4 ones and write 14. Why? Because we traded 1 ten for 10 ones. We already had 4 ones, and we know that 4 ones + 10 ones = 14. Good explanation. Are we able to subtract now? Yes, we need to subtract 5 from 14, which equals 9.

Step Three: Go to the tens. What does it say? Do you need to trade? If no, write down the difference below the tens column. If yes, record taking a hundred and trading it for 10 tens. Write down the remaining number of hundreds and how many tens you now have. After trading, subtract the tens and record the difference in the tens column.

$$
\begin{array}{r}
\overset{12}{} \\
4\,\cancel{2}\,14 \\
\cancel{534} \\
-\;275 \\
\hline
59
\end{array}
$$

Step Four: Go to the hundreds. What does it say? Do you need to trade? If no, write down the difference below the hundreds column. If yes, record taking a thousand and trading it for 10 hundreds. Write down the remaining number of thousands and how many hundreds you now have. After trading, subtract the hundreds and record the difference in the hundreds column. Continue with other place values if necessary.

$$
\begin{array}{r}
\overset{12}{} \\
4\,\cancel{2}\,14 \\
\cancel{534} \\
-\;275 \\
\hline
259
\end{array}
$$

Step Three: What do we do next? Visualize what we did with disks and drawings. Pretend you are using disks if you need to. Since we have finished the ones, we need to move next door to the tens place. What does it say in the tens? It says 2 tens minus 7 tens. Are we able to do that easily? No, so we need to trade. Why? We need more tens. Where are we going to get tens from? The hundreds. Okay, so we need to trade in 1 hundred for 10 tens. How will we record that? Think about what you just said. You said we needed to trade in a hundred, so cross out the 5 in the hundreds place. Why? Because you're trading in 1 hundred for 10 tens. How many hundreds will be left? 4. Record the 4 above the 5 hundreds. What did we do with the hundred? We traded it for 10 tens. So, should I record a 10 above the 2? No! You already had 2 tens, which means you now have 12 tens. Cross out the 2 tens and write 12 above it. Now what? We need to subtract 7 tens from 12 tens, which equals 5 tens.

Step Four: Now we're done! No? We still have to do the hundreds place. What does it say in the hundreds? It says subtract 2 hundreds from 4 hundreds. Are we able to do that easily? Yes. Okay, go ahead and do it. What is 4 – 2? 2. Is it really a 2? No, it's really a 200, but it looks like a 2 because the 2 is in the hundreds place. Now we've found the difference. The difference between 534 and 275 is 259.

Multiplication

MULTIPLICATION IS AN EXCITING CONCEPT for young mathematicians. Being able to multiply makes them feel like "big kids." The temptation, for both kids and parents, is to jump right into memorizing multiplication facts. If you see this happening, intervene. Before trying to memorize facts, students need a real understanding of multiplication. They must internalize the concept that multiplication is repeated addition. They have been doing it for years as they skip counted by 1, 2, 5, and 10. They just weren't calling it multiplication then.

As they learn strategies for multiplication, students will be asked to decompose numbers, sometimes into tens and ones and other times into factors and products that are familiar or easy for them. Knowing the concept of part-whole facilitates this ability. It also makes using multiplication strategies to solve word problems much easier.

You'll notice that the sequence for teaching multiplication strategies follows the concrete-pictorial-abstract (C-P-A) approach too. Most of the strategies, starting with number bonds, have a concrete component, such as counters or place value disks, while others use a visual aid such as place value strips. The C-P-A approach will be extremely important when you begin to teach the traditional method of multiplication. Pay close attention to the use of groups of place value disks to show the concept. Students will then be able to experience expressions such as 4 x 26 as 4 groups of 2 tens and 6 ones. From there, have them draw pictures, and then graduate to solving abstractly.

Making connections between multiplication strategies is vital. Students need a clear understanding that learning to use number bonds and to decompose numbers is both necessary and helpful for the distributive property, area, and place value chart methods. They also must recognize that all of these methods are really the same thing: repeated addition that pays attention to place value. One way to help students see the connections is to use similar manipulatives. Thus, if you use place value strips as a manipulative to demonstrate one strategy, also show how you would use the strips in working with another strategy. Guide students to discover these connections by asking them, "What are you noticing about these two methods?"

After working with the distributive method for multiplication, you'll begin teaching the area model. In essence, these two strategies show similar things—the importance of place value with multiplication—but they are visually different. Use left-to-right to compute a multiplication problem such as 6 x 84. Then compute the same problem using the area model. As you do, ask your students, "What's the same?

What's different?" Encourage students to notice and share the connections between the two methods.

Problem-solving is the heart of mathematics. We teach our students multiplication strategies so that they can apply them to word problems where equal groups are being put together, not simply to solve pages of multiplication equations. A daily problem-solving session is a great time to encourage your students to use all the strategies they've learned. You'll also want to encourage your students to write their own multiplication math stories. If they truly understand the concept of multiplication, this task will be quite easy.

Stages of Multiplication

Mastery of multiplication proceeds from the basic concept of multiplication as repeated addition to learning facts through number bonds and then progresses to multiplying multidigit numbers, fractions, decimals, and more. Below is a recommended sequence for teaching the concept of multiplication, starting with the concrete and ending with the abstract:

1. Number bonds

2. Place value disks and charts

3. The distributive property

4. Area model

5. Traditional multiplication

Number Bonds

STUDENTS NEED A SOLID UNDERSTANDING that multiplication is about putting together equal groups. This applies whether they are working on basic facts or multi-digit multiplication. Starting with number bonds when you teach multiplication offers students a much better conceptual foundation than simply memorizing multiplication facts. Just as students learned that number bonds show the part-whole relationship for addition, they will discover that number bonds can show the factor-product relationship for multiplication. Number bonds also help students to see the relationship between multiplication and division. For example, once they've figured out the number bond relationship of 8, 6, and 48 for multiplication, they'll also be able to work out the relationship of those numbers for division.

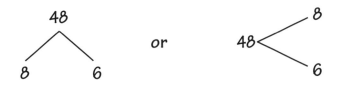

It's important to vary your vocabulary when you talk with your students about number bonds. Use the terms equal groups, factors, and number bonds frequently and interchangeably so that students know there is more than one way to say the same thing. You'll notice that I do this later in the guided conversation. As you demonstrate number bonds for multiplication, it's a good sign if your kids say something like, "That's the same way we wrote number bonds for addition and subtraction." They're noticing patterns and relationships while making connections with previous learning!

INTRODUCING THE STRATEGY

Step One: Select a number (the product) and write it down. Make sure the number you select is a composite number, that is, a number with more than just 1 and itself as factors.

12

Step Two: Think of different ways to break up the number into factors. What are two factors (parts) that will go together to make the product?

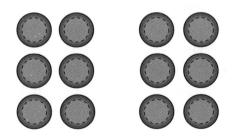

Guided Conversation

Step One: Today we're going to explore different ways to break numbers into groups. We'll call these number bonds. Remember when we made number bonds for addition and subtraction? Now we're going to make them for multiplication. The difference with multiplication is that we will be thinking of groups and how many will be in each group. We'll call these groups and their sizes factors. The total will be called a product. Later, we'll use this concept of factors and products to help us with division too. Right now, though, I can hardly wait for you to see this. Are you ready? We're going to start with the number 12.

Step Two: I want to think of some different ways to break the number 12 into groups, or factors. One thing we need to know is that the groups must be equal in size. Let's get out 12 counters. What is one way we can divide these 12 counters into groups? Remember, the groups have to be equal in size. I'm thinking of splitting the 12 counters into 2 groups. Take your counters and split them into 2 equal groups. How many groups? Yes, 2 groups. Go ahead. How many counters are in each group? Each group should have 6. If we were talking about addition, we would say 6 + 6 = 12, but since our topic today is multiplication, we are going to say we have 2 groups of 6. Here's how we write that: 2 x 6 = 12. So we say that 2 and 6 are factors of 12. What do we have? That's right. We have 2 groups of 6 for a total of 12. That means 2, 6, and 12 are a number bond for multiplication.

Step Three: Find other ways to break up, or decompose, the product into two factors. Are there only two factors? How many can you find?

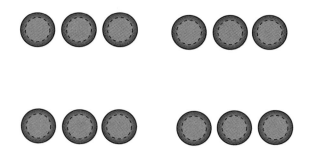

Step Four: Record the number bonds that you found. Notice patterns when looking for factors, and share what you discover with your partner, group, or class. (You may also want to model that number bonds can be written horizontally.)

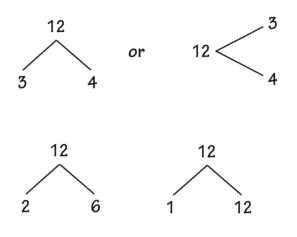

Step Three: Let's put the 12 counters back together into one group. Wait! Could this one group be a factor? Yes! If you have 12 counters, you could have 1 group of 12. It's like buying one box of a dozen donuts. In other words, 1 and 12 are factors of 12. That means 1, 12, and 12 are a number bond for multiplication.

How about if we try 4 groups? Use your counters to see if we can make 4 equal groups. Can we? Yes! How many will be in each group? 3. Oh! So there are 4 groups of 3. 4 and 3 are factors of 12 too. That means 4, 3, and 12 are a number bond for multiplication. I wonder if there are more. Wow! You found 3 groups of 4. Is that the same as 4 groups of 3? Show how they are different with your counters. They have the same product and the same factors, but the picture doesn't look quite the same with our counters.

Step Four: What was one way we could make factors of 12? That's right. 3 and 4 are factors of 12. Let's prove it one more time with our counters. Good. Now let's act out this number bond. Watch me and copy my actions. Put out one hand and say, "We have 3 groups." Put out your other hand and say, "Each group has 4 in it." Put your hands together over your head and say, "The product is 12." Try it again. 3 (one hand out) and 4 (other hand out) are factors of 12 (hands over head).

Remember how we wrote number bonds for addition and subtraction? Watch how I do the same thing for multiplication. Look how I write it. Now you write these factors and product as a number bond for multiplication.

With a partner, would you please show a different way to make equal groups with your 12 counters? Act it out, saying the factors and the product. Then write down your new number bond. Can you find any other ways to make equal groups that have a product of 12? Good luck!

Students need to practice the use of number bonds and the language that's associated with them, including part-whole and factor-product. Have them start with easy combinations like these:

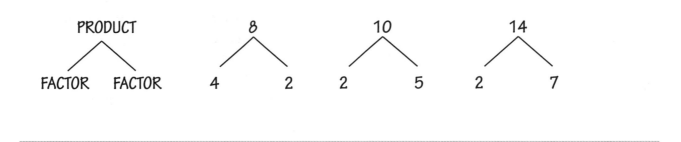

Next, move on to more challenging examples and mix things up a bit. Give your students the factors and have them figure out the product. After that, introduce bonds using numbers with greater values. Then present a bond with only one factor and the product shown and ask your students to figure out the missing factor.

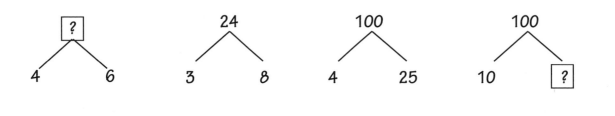

Multiplication of decimals, fractions, units of measure, and time can also be shown as number bonds.

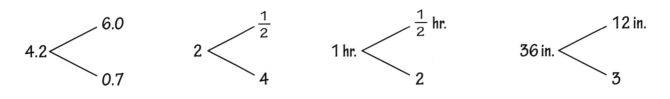

Place Value Disks and Charts

USING PLACE VALUE DISKS IN MULTIPLICATION helps to build understanding because it literally shows students what's happening when they perform regrouping. They don't just follow a procedural rule; they learn why regrouping needs to occur and how to record their multiplication on paper.

As I've mentioned before, when your students first begin manipulating disks to solve multiplication problems, don't ask them to record their work on paper. While the students use their disks, you may want to record what they are doing as a way of modeling for your students. Then, when you feel your students understand the concept with disks, ask them to both manipulate the disks and record on paper with disk drawings. Once students understand that they are drawing exactly the same steps that they are performing with their disks, they can stop using the disks.

For the next stage in the process, ask your students to draw disks to show what's happening and also record abstractly. Make sure they draw and then record what they've done abstractly one step at a time. In other words, they should not do all of their drawing first and the abstract recording afterward. Your goals are for kids to understand why they are writing numbers in certain locations when they record and also for them to see the connection between the disks, drawings, and the abstract representation.

This is a great opportunity for students to work with partners. One partner may draw the disks while the other student writes the abstract expression. Or one student may become the explainer while the other is the listener to make sure that all steps are explained correctly.

Be wise with the numbers you select. Remember that the point of using the disks is for kids to "see" what's happening with the multiplication and eventually be able to connect it to the traditional algorithm. Devise problems that require trades to be made (such as 3 x 53, 4 x 26, or 4 x 135), but avoid numbers that would require so much disk work that finding a solution will be time-consuming and awkward (examples: 97 x 8, 39 x 7, or 46 x 6).

INTRODUCING THE STRATEGY
One-digit by three-digit multiplication

Step One: Write a multiplication
expression. Think about what the equation means. If it's helpful, use the commutative property to change the order of the numbers.

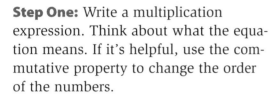

$$126 \times 4$$
$$4 \times 126$$

Step Two: Look at the first number in the expression. This number tells you how many rows you'll need to make. Draw that number of rows on your place value chart.

Step Three: Look at the second number in the multiplication sentence. The second number tells you how many are in each group or row. Use disks to build this number in each row.

(100)	(10) (10)	(1)(1)(1)(1)(1)	(1)
(100)	(10) (10)	(1)(1)(1)(1)(1)	(1)
(100)	(10) (10)	(1)(1)(1)(1)(1)	(1)
(100)	(10) (10)	(1)(1)(1)(1)(1)	(1)

Guided Conversation

Step One: Today we're going to use place value disks to try solving a three-digit by one-digit multiplication problem. Write down 126 x 4. What does 126 x 4 mean? It means there are 126 groups of 4. Would you like to build 126 groups? No! That's a lot of groups. Can you think of what we could do to make this easier? Yes, we can change it to 4 x 126. Now what does it mean? 4 groups of 126. Is this the same thing? Yes. Why? We just did a switcheroo on it. The math term for that is the commutative property.

Step Two: Because there are 4 groups, we need to draw 4 rows. How many rows do we need to make? 4. Draw or make your 4 rows.

Step Three: The second number in our expression tells us how many are in each group. That means we have to build what number? 126. Think: how do we build 126? Yes, 1 hundred, 2 tens, and 6 ones. How many 126s do we need to build? 4. How do you know? The equation said 4 x 126, which means 4 groups of 126. Double-check that you've built the correct number in each row.

Step Four: Put the ones together. Are there enough ones to trade or regroup? If so, make the trade. (Make sure students verbalize what they are doing as they regroup or trade.)

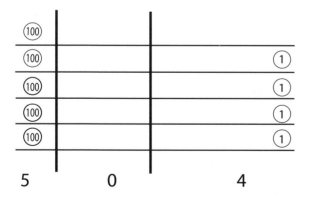

Step Five: Put the tens together. Are there enough tens to trade or regroup? If so, make the trade.

Step Six: Put the hundreds together. Are there enough hundreds to trade or regroup? If so, make the trade. If not, record the product.

5   0   4

4 x 126 = 504

Step Four: Now we're going to add, or we could skip count, all of the ones. Put the ones together. How many ones do we have? 24. Do we have enough ones to make a trade? Yes, because we have more than 10. What should we do? Take 10 ones and trade them for a ten. Do we have enough to make another trade? Yes. What should we do? You got it. Trade them for another ten. Do we have enough to make a third ten? No, we have only 4 ones left. So how many tens did we end up trading our ones for? 2. Why? Because we had 24 ones and that's the same as 2 tens with 4 ones left over.

Step Five: We've finished the ones, so where should we go? The tens. Add the tens together. How many tens do we have after we put them together? 10. How did we get 10 tens? We added the 2 tens from each of the 4 rows and then added the 2 tens we got when we traded in our ones. Do we have enough tens to make a trade? Yes. Why? We have at least 10. What should we do? That's right; we should take 10 tens and trade them for a hundred. Do we have enough to trade for another hundred? No. Why? We have no more tens.

Step Six: What should we do now? Put the hundreds together. Do we have enough hundreds to make a trade? No. Why not? We have only 5 hundreds. How many would we need to make a trade? We'd need at least a total of 10 hundreds to trade. If we had enough, what would we trade for? Yes indeed, a thousand.

What do we have now? 5 hundreds, 0 tens, and 4 ones. How much is that? 504. Let's record that. Write the equation 4 x 126 = 504.

Step One: Write a multiplication expression. Think about what it means. If it's helpful, use the commutative property to change the order of the numbers.

6 × 23

Step Two: Look at the first number in the expression. This number tells you how many rows you'll need to make. Draw that number of rows on your place value chart. Now look at the second factor. Draw disks in each row to match the second factor.

Step Three: Put the ones together. Are there enough ones to trade or regroup? If so, show the trade by circling the 10 ones and drawing the ten you gained in the tens place.

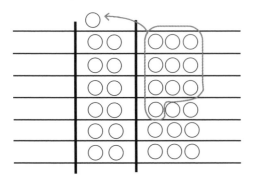

Guided Conversation

Step One: We've been learning about multiplication by working with our place value disks. Now we're going to use what we know and draw the disks instead. You can pretend that your disks are right in front of you if it helps. Write the multiplication sentence 6 x 23. What does 6 x 23 mean? It means there are 6 groups of 23.

Step Two: Now we are going to draw the place value chart and disks. We'll start by drawing the place value chart. How many rows do we need? You're right, 6. Draw 6 rows. What's the second factor? 23. Think about how we built 23 with place value disks. How do you think we're going to draw it? You got it! We're going to do exactly the same thing we would have done with disks. We'll draw 2 tens and 3 ones. How many times do we need to draw 23? 6. Why? Because our expression says 6 x 23, which means 6 groups of 23.

Step Three: Now that we've made a picture of 6 x 23, what do you think we should do? Think about what we did when we had the disks out. Yes, we put the ones together. Do we have enough to make a trade? Yes. Why? We have at least 10 ones. What should we do? Trade in 10 ones for a ten. How do we record that on our paper? Circle the ones to show that you are trading them for a ten. Then draw an arrow to show where the new ten you just traded for belongs. (Make sure students verbalize what they are doing as they regroup or trade.) Do we have enough ones to make another ten? No, we have only 8 ones left.

Step Four: Put the tens together. Are there enough tens to trade or regroup? If so, draw the trade by circling the 10 tens and trading for a hundred. Draw the hundred in the hundreds place.

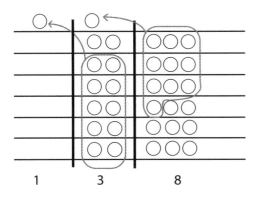

1   3   8

Step Five: Are there enough hundreds to trade or regroup? If so, draw the trade by circling the 10 hundreds and trading in for a thousand. Draw the thousand in the thousands place. If there are not enough hundreds to make a trade, simply write out the completed equation.

$6 \times 23 = 138$

Step Four: We've finished the ones. What should we do now? That's right; put the tens together. Do we have enough to make a trade? Yes. Why? We have at least 10 tens. What should we do? Trade for a hundred. What do we exchange for the 10 tens? Correct. 1 hundreds disk. How do we record that on our paper? Circle the 10 tens. Draw an arrow to the hundreds place, and draw a hundreds disk to show your trade.

Step Five: Do we need to do anything in the hundreds? No, because we have only 1 hundred. What do we have left to do? We must record the complete multiplication sentence: 6 x 23 = 138.

The place value method also works well for decimals. Instead of using just place value disks, get out place value decimal tiles, put a decimal point on your place value chart and follow the same explanation process we used previously. Of course, the difference is that you regroup 10 tenths for a ones disk or 10 hundredths for a tenth. As you present, you'll need to emphasize language and be sure to clearly pronounce the "ths" in tenths and hundredths.

4 x 3.5

Step 1

Step 2

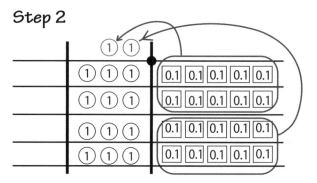

Step 3

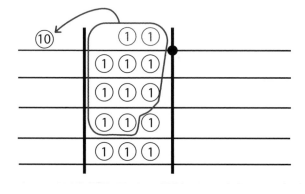

Step 4

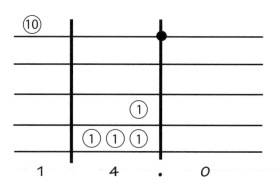

4 x 3.5 = 14.0

The Distributive Property

AFTER STUDENTS UNDERSTAND THE CONCEPT of making equal groups in multiplication and have practiced using place value disks to reinforce the concept, you can introduce the distributive property strategy.

Using the distributive property strategy to teach the concept of multiplication is similar to the left-to-right strategy for teaching addition because of the strong emphasis on place value. You will take each number and pull it apart by place value. So, for example, 456 will become 400, 50, and 6. This is easily modeled with place value strips, so that students have a visual reminder of the value of each place as they begin to multiply each part using the distributive property.

If your students are struggling with multiplication of numbers in the tens or hundreds, try replacing numbers with words. For example: 3 x 50 becomes 3 x 5 tens, which is equivalent to 15 tens. Ask your students something like this: "What do we know about tens? They have one zero. So 15 tens = 150." Similarly, 4 x 700 becomes 4 x 7 hundreds. That's 28 hundreds, which equal 2,800. Since place value is the emphasis with this method, it also works well for decimals!

You may find that as students solve problems using the distributive method, they will add the partial products in different ways. Ask students to share their various methods. What did they do? Why did they do it that way? Who did it a different way? Celebrate the many different methods students use to find the same answer.

After your students have mastered the distributive property for multiplication on paper, challenge them to try using the distributive method mentally instead of recording all of the individual steps on paper.

INTRODUCING THE STRATEGY

Step One: Write a multiplication expression horizontally.

$$36 \times 4$$

Step Two: Decompose each number by its place value. When you begin, you may want to use place value strips to "prove" the values. Write down the new expanded multiplication expression.

$$36 \times 4$$
$$(30 \times 4) + (6 \times 4)$$

Step Three: Multiply the single factor by the tens. Record the product.

$$36 \times 4$$
$$(30 \times 4) + (6 \times 4)$$
$$120$$

Step One: We've been working on number bonds and using place value disks for multiplication, and today we're going to learn yet another strategy. First, I'm going to write a multiplication expression horizontally. Remember that means the expression is "lying down," or written on one line straight across.

Step Two: Next, I need to decompose the numbers, like we did when we were studying addition. In this expression, can I decompose both of the numbers? No, that's right; I can only decompose the two-digit number, 36. I'm going to begin with the greatest value, which is the tens. In the number 36, I see 3 tens, which I know is 30. So I need to write 30 x 4 because I have to multiply each part of 36 by 4. Next, I need to look at the ones. I see 6 ones, so I'm going to write 6 x 4. Remember to put parentheses around each set of numbers that need to be multiplied.

Step Three: I know that I need to multiply each set of numbers inside the parentheses. We'll start with the largest value. In this case, the largest value is the tens. I see that I need to multiply 30 x 4. I know that 3 tens x 4 = 12 tens. Tens have one zero, so 30 x 4 = 120. So I'm going to write 120 underneath 30 x 4.

Step Four: Multiply the single factor by the ones. Record the product.

36 × 4

(30 × 4) + (6 × 4)

   120 + 24

Step Five: Find the sum of the partial products.

36 × 4

(30 × 4) + (6 × 4)

   120 + 24 = 144

Step Four: After the tens come the ones. What do you think I'm going to multiply now? Yes, I'm going to multiply 6 x 4. 6 x 4 = 24. Then I'll write 24 underneath 6 x 4. We've now found the products of the two different parts. These are called partial products. Next, make sure that you put a plus sign in between the 120 and the 24 because we have to put the partial products back together.

Step Five: Whew! We're getting close to finding the product of 36 and 4. All we have to do is add the partial products together. I can add the numbers on paper, or I can try to do it mentally. I'm going to put 120 in my mind. I see I need to add 24. First, I'm going to add 20, or 2 tens, to 120. That gives me 140. Now I'll add the remaining 4 ones for a final total, or product, of 144. Hey! I just used the left-to-right mental math method that we learned earlier this year to find the answer.

The distributive property strategy can be used with multidigit multiplication as well as decimals. If the problem involves numbers in the hundreds or thousands, explain to students that they should start with the largest value and continue from there. Below are several examples that deal with two-, three-, and four-digit multiplication as well as decimals.

After your students are feeling confident with two-digit multiplication, give them a three-digit problem. You'll want to work together with your students to compute problems similar to the ones below and then offer time for independent practice. Remind your students to keep the commutative property in mind as they tackle multidigit multiplication. Just as students may prefer to work with 9 x 5 instead of 5 x 9, the same will be true for multidigit numbers. For example, for many students it will be easier to use the commutative property to change 673 x 8 to 8 x 673.

67 x 8 (or, use the commutative property to solve as 8 x 67)

(8 x 60) + (8 x 7)

480 + 56 = 536

---

9 x 542

(9 x 500) + (9 x 40) + (9 x 2)

4,500 + 360 + 18 = 4,878

2,308 x 7 (or, use the commutative property to solve as 7 x 2,308)

(7 x 2,000) + (7 x 300) + (7 x 8)

14,000 + 2,100 + 56 = 16,156

---

8 x 6.7

(8 x 6) + (8 x 0.7)

48 + 5.6 = 53.6

Area Model

THE AREA MODEL IS A GREAT VISUAL MODEL for multiplication because it draws attention to place value. The area model requires students to draw a rectangular visual for each expression. If the expression is two-digit by two-digit, then a two-digit by two-digit array will be drawn. Each rectangle or box in the array represents a place value. As you teach the area model, encourage students to notice the similarities between the area model and the distributive property strategy. Ultimately, the result achieved for any given multiplication problem will be the same with the area model as with the distributive property, but the presentation will look different because of the rectangular nature of the area model. Students who have difficulty with drawing and organization will find graph paper is very helpful when they are solving problems using the area model.

When students have finished their area models and are finding the totals of the partial products, they are likely to use many different approaches to finding the answer, including using the associative property to change the groupings being added. Celebrate this diversity by asking your students to share what they did, and praise them for their success. See how many different ways students have used to find the final answer. And, as always, encourage them to use correct math terms as they explain their work.

If you are studying the concept of area with measurement and geometry, you could also include a lesson on the formula Area = Length x Width.

INTRODUCING THE STRATEGY
One-digit by two-digit multiplication

Step One: Write down the multiplication problem you need to solve. Draw squares to represent the number of "places" in each number. In this case, you will draw two squares horizontally next to each other (I call this a 1 by 2 array).

7 x 42

Step Two: Write the expanded factors above and beside the squares in the array. The placement of each number corresponds to its place value.

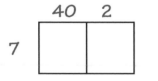

Step One: We're going to learn a new multiplication strategy today called the area model. It is related to the distributive property. Let's get started. We need to decide what multiplication problem we are going to solve. Let's try 7 x 42. Write down 7 x 42. We will need to decompose each number by its place value, but first we're going to draw squares for each place value. Since 7 is a one-digit number (ones only) and 42 is a two-digit number (and thus has ones and tens), we consider this a one-digit by two-digit expression, so we'll draw a 1 by 2 array with squares. Watch what I do. I'm going to draw one square and then I'm going to draw one more square connected to it. Now you do it. Why did I do that? Yes, I did it because it's a one-digit by two-digit multiplication expression.

Step Two: Now we'll write the decomposed numbers outside of the squares. What do we get when we decompose 42? Yes, 40 and 2 because 40 and 2 makes 42. Notice how I wrote 40 above the first square and 2 above the other square. Go ahead and do it yourself. Neighbors, check to make sure your partner has written the 40 and the 2 in the correct locations. Now I'll write 7 on the left side of the first square. Now you do it. Partners, check your neighbor's work again please.

Step Three: Multiply the decomposed numbers. Write each partial product in the corresponding square.

$$
\begin{array}{c|c|c}
 & 40 & 2 \\
\hline
7 & 280 & 14 \\
\end{array}
$$

Step Four: Find the sum of the numbers inside the squares, and write out the answer as a horizontal equation.

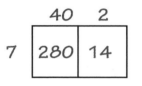

$$7 \times 42 = 280 + 14$$

$$280 + 14 = 294$$

$$7 \times 42 = 294$$

Step Three: We've finished setting up the array and now we're ready to multiply. We are going to multiply the numbers on the outside of the squares in a way that's similar to the way we use a multiplication table. To figure out which two numbers to multiply together, go straight across from the first factor and straight up to find the second factor. We'll write the products inside the squares. First, we'll multiply 7 and 40. What is 7 x 40? That's right; it's 7 x 4 tens, which equals 28 tens, or 280, so 7 x 40 = 280. Write 280 inside the square next to the 7 and below the 40. Now we need to multiply 7 and 2. What is 7 x 2? Yes indeed, it's 14. I'll write 14 in the square under the 2. Be careful to put the partial product in the correct square by going straight across from one factor and straight down from the other factor.

Step Four: Now that we've multiplied, we need to find the sum of the numbers inside the squares. We can do this! It's the same left-to-right mental math addition strategy that we've been using in our class for a long time. What numbers do we need to add? We need to add 280 and 14 because they are the partial products. In your mind, think: what's 280 + 14? Pay attention to place value. We can start by thinking 280 + 10 = 290. Then I need to add the other 4. 290 + 4 = 294.

Rewrite the original problem, followed by the partial products. So, when I look at your work, I will see your area model drawn, but I also will see 7 x 42 = 280 + 14. The sum of 280 + 14 = 294, so the product of 7 x 42 is 294. When you write out your work this way, you can see all the stages you went through to find the answer.

INTRODUCING THE STRATEGY
Two-digit by two-digit multiplication

Step One: Write down the multiplication problem you need to compute. Draw squares to represent the number of "places" in each number. In this case, you will draw a 2 by 2 array.

17×34

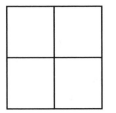

Step Two: Write the expanded factors above and beside the squares in the array in the appropriate spots to correspond with their place values.

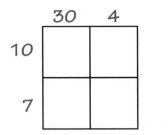

Guided Conversation

Step One: We've practiced using the area model for one-digit by two-digit multiplication. Now we're going to try two-digit by two-digit multiplication. Do you think we can do it? Yes! Do you think it's going to be hard? No. Why? Because it's just the same, but with an extra row of squares. Yes, the array will be a little bit bigger than it was for the one-digit by two-digit problem. How did we start before? Oh, that's right. We decomposed the numbers by place value and then drew squares to correspond to place value. I remember that we drew a 1 by 2 array for our one-digit by two-digit multiplication problem. What do you think we're going to draw for this problem? You're absolutely correct. Since our problem is two-digit by two-digit, we'll draw a 2 by 2 array.

Step Two: Now we'll write the decomposed factors in the right spots above and beside the squares. What will we decompose 34 into? Yes, 30 and 4. Where should I write the 30 and the 4? That's right. The 30 will be above the first column, and the 4 will be above the second. Each gets its own space. What about 17? How should I decompose it? Good, 10 and 7. Where should I write the 10 and 7? Oh, that's correct; along the left side of the array. I'll write 10 next to the upper row and 7 next to the lower row.

Step Three: Multiply the decomposed numbers. Write the partial products inside the corresponding squares.

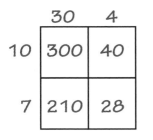

Step Four: Find the sum of the numbers in the squares, and write out the answer as a horizontal equation.

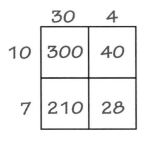

$$17 \times 34 = (300 + 210) + (40 + 28)$$
$$510 \quad + \quad 68 \quad = 578$$

$$17 \times 34 = 578$$

Step Three: Let's think back to our other area model practice. After we write the expanded numbers outside of the array we need to multiply to find the partial products. What should we begin with? Good idea. We'll begin with 10 x 30. That's 1 ten x 3 tens, which is equivalent to 3 hundreds. Where should I write 300? Inside the top left square. Why there? Because that's where the 10 and the 30 meet. What should I multiply next? I'm going to multiply 10 and 4. 10 x 4 = 40. I'll put the product 40 in the square immediately below the 4. What's next? Yes, I need to multiply 7 and 30. That's the same as 7 x 3 tens, which equals 21 tens, which equals 210. I'll write that in the square where the 7 and 30 meet in the array. Is this the correct space to write in? Yes. How do you know? It's where the 7 and 30 meet if I run my finger straight down from the 30 and straight over from the 7. Finally, I need to multiply 7 and 4. 7 x 4 = 28, and I'll write that partial product inside the remaining square.

Step Four: Are we done? No! We've done all of the multiplying, but we need to find the sum of the numbers inside of the squares. What do we call these numbers? That's right; they are partial products. Let's write down what we've done. First, we need to write 17 x 34 since that's the multiplication expression we began with. Then we need to write down all of the partial products. Does it matter what order we write them in? No! The commutative property for addition allows us to add in any order. If you decide to add the numbers in a different order than your neighbor does, you should still get the same answer. For this example I'm going to start with the largest partial product first. So, after 17 x 34, I'll write = 300 + 210 + 40 + 28. Finally, I'll add my partial products. I'm going to start with the 300 and add 210 since they are both in the hundreds. 300 + 210 = 510. Next, I'll add 40 + 28 = 68. Now it's time to put it all back together. I have 510 + 68. We can write down these parts if it will help, or we can try to do it mentally. Let's see, 510 + 60 = 570, and if I add another 8, it'll be a total of 578.

Building On

One of the things I like best about the area model is its versatility with numbers of different sizes. It can be used to multiply two-digit by three-digit numbers to emphasize place value, to multiply three-digit by four-digit numbers, and even to multiply two-digit decimals by two-digit decimals. If you use the area model for decimals, remember to check student work to make sure they are placing the decimal point in the correct position. And keep in mind that while the area model can be used for problems involving numbers into the thousands and millions, it may not be the most efficient strategy.

3×234

	200	30	4
3	600	90	12

$600 + 90 + 12 = 702$

$3 \times 234 = 702$

38×5.5

	5	0.5
30	150	15.0
8	40	4.0

$(150 + 40) + (15.0 + 4.0)$

$190 \quad + \quad 19.0 \quad = 209$

$38 \times 5.5 = 209$

23×717

	700	10	7
20	14,000	200	140
3	2,100	30	21

$(14,000 + 2,100) + (200 + 140) + (30 + 21)$

$16,100 \quad + \quad 340 \quad + \quad 51$

$16,100 \quad + \quad\quad 391 \quad\quad = 16,491$

$23 \times 717 = 16,491$

Traditional Multiplication

THIS IS THE METHOD MOST OF US LEARNED GROWING UP. When teaching the traditional multiplication algorithm, connect what is happening in the computation with other methods. Ask your students to pretend to manipulate disks as they work. Students should be asking themselves: "What's the next step? Do I need to trade? Why or why not?"

Giving students opportunities to work with partners and communicate mathematically about what they are doing and why helps in developing the multiplication algorithm concept. Drawing pictures before solving equations using the traditional algorithm alone will also help students make the connections among the disks, the drawings, and the abstract computation.

I strongly suggest working with your students to make a couple of charts that show drawings of disks alongside the abstract algorithm. If you draw each step in a separate color (circling ones and making a trade for a ten using blue ink, circling tens and making a trade for a hundred using green ink, and so on), this will help students visualize what they are doing and why.

When or if you ask your students to try two-digit by two-digit or two-digit by three-digit multiplication using the standard algorithm, make sure that they can explain their work, and that they are not just following rote rules. Of course, the same is true when applying the traditional algorithm to decimals.

INTRODUCING THE STRATEGY

Guided Conversation

Step One: Select a multiplication problem to solve. Write the problem vertically.

$$\begin{array}{r} 89 \\ \times\ 6 \\ \hline \end{array}$$

Step Two: Begin by multiplying the ones. Visualize what you did with the disks as you are working. Record any trades you need to make in the tens place and any leftover ones below the line in the ones place.

$$\begin{array}{r} {}^5\ \\ 89 \\ \times\ 6 \\ \hline 4 \end{array}$$

Step Three: Look at the tens. Multiply the tens in the top number by the bottom number. Add the traded tens to find a tens total. Again, think about what you did with disks. Record any trades you need to make in the hundreds place and any leftover tens below the line in the tens place.

$$\begin{array}{r} {}^{5\,5}\ \\ 89 \\ \times\ 6 \\ \hline 34 \end{array}$$

Step One: We've been learning about multiplication using disks and drawings and arrays. Today we are going to think about disks and drawings while we are recording our work using numbers only. We'll start with 89 x 6. Write this expression vertically. What does 89 x 6 mean? It means there are 89 groups of 6.

Step Two: Let's begin. Start in the ones. What does it say? 6 x 9. What is 6 x 9? 54. Correct. Do we have enough to trade? Yes. How many trades? 5. How do you know? Because 54 has 5 tens. Where do we record this trade? In the tens. What should I write down? A 5 above the tens place. How many ones are left? 4. Where should I write these? Below the line in the ones place.

Step Three: Go to the tens. What does it say? It says 6 x 8, but we know that it's really 6 x 8 tens, which equals 48 tens. Do we have enough tens to trade? Yes. There's something tricky I need to do first. Before I trade, I have to add the 5 tens that we traded earlier. 48 tens + 5 tens = 53 tens, which also equals 530. How many trades? 5. Why? Because 530 has 5 hundreds. Where do we record this trade? Above the hundreds. What should I write down? A 5. Where did we get the 5? Oh yes. We multiplied 6 x 80 and got 480, and then we had to add the 50 that was traded earlier. If I make 5 trades to the hundreds, how much is left? 30, or 3 tens. Where should I record that? In the tens place. How should I record it? Just write a 3 below the line in the tens place.

Step Four: Look at the hundreds. Multiply the hundreds (in this case there are zero) by the bottom number. Add the traded hundreds to get a hundreds total. Record the hundreds in the hundreds column.

$$
\begin{array}{r}
\scriptstyle 5\ 5 \\
89 \\
\times\quad 6 \\
\hline
534
\end{array}
$$

Step Four: Go to the hundreds. There are no hundreds to multiply, but I have 5 hundreds that I traded. So I need to record that 5 in the hundreds place. I'll write it below the line in the hundreds place. Now I have gone through all of the steps just like I did when I was using the place value disks. The answer to our problem is 89 x 6 = 534.

Division

ALTHOUGH CHILDREN HAVE AN IMPRESSION OF DIVISION as more advanced than, say, addition, it's just as much a part of their day-to-day experiences as addition, subtraction, and multiplication. Think of kindergarteners passing out birthday treats or first graders sharing a bag of pretzels or a pizza. This is division in context. These are applications of the division concept.

So, if children begin using division very early in life, why is it such a difficult concept for so many of our students? My answer is that children often view division only as a set of number facts and long-division rules to memorize. They don't see the connection between real life and math class. This heightens the challenge of teaching division.

What can we do differently to help our students master division? This is a familiar theme by now, but the most important way we can help is by developing the division concept first. Help students recognize that they use division regularly by pointing out examples. Focus first on the concept of sharing and putting items into equal groups. Begin by using manipulatives to develop an understanding of basic division facts, such as $20 \div 5$. Just as multiplication is repeated addition, kids need to grasp that division is repeated subtraction.

Once students have mastered basic division facts, it's time to move on and teach other division strategies. Make sure you model and think aloud for your students. This includes thinking aloud about using multiplication facts to help with division. Tell students what you are doing and why. Have students listen for the language that you are using as you explain a strategy. When students are able to build division representations with counters, such as bingo chips, and explain what they are doing, then it's time for them to draw (pictorial). The final step in the teaching sequence is writing and solving equations abstractly.

One big challenge for teachers when they introduce division is that division can be seen in two different ways: partitive division and measurement division (also called quotitive division). Thus, the same division equation can mean two different things. Take the equation $32 \div 4 = 8$. With partitive division, $32 \div 4$ would be interpreted as taking 32 items and putting them in 4 groups. On the other hand, if you view the problem as measurement division, $32 \div 4$ would be interpreted as 32 items being put into groups of 4. Think about your teaching.

Which type of division do you model most often? I'm willing to bet it's partitive. Challenge yourself to use both.

Stages of Division

Begin by building the concept of division as repeated subtraction. Through the use of number bonds, you can demonstrate the inverse relationship between multiplication and division—one is the undoing of the other. Then you can move on to long division of multidigit numbers and decimals and even making connections to fractions. The following is a recommended sequence for teaching the concept of division:

1. Number bonds

2. Place value disks and charts

3. The distributive property

4. Partial quotient division

5. Traditional long division

6. Short division

Number Bonds

WE'VE USED NUMBER BONDS TO INTRODUCE addition, subtraction, and multiplication, and now we'll use them again for division. The concept is the same—helping students see relationships between numbers. In multiplication, we focused on finding the product of two factors. Now we'll work to find one factor, when we already know the total (the dividend) and the other factor.

If students have mastered number bonds in multiplication, then division is going to be a "piece of cake." Instead of having to memorize, for example, $56 \div 7 = 8$, $56 \div 8 = 7$, $7 \times 8 = 56$, and $8 \times 7 = 56$, students learn that 56, 7, and 8 are related to one another. My goal is this: when a student is presented with 56 as the total, or dividend, along with the factor 8, he will immediately say something like, "No problem. I know that $8 \times 7 = 56$, so that means 7 must be the missing factor."

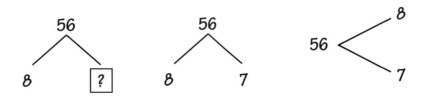

You may want to work with your students to prepare a chart with multiplication and division vocabulary terms, like the one shown below.

Factor x Factor = Product

$7 \times 8 = 56$

Dividend ÷ Divisor = Quotient

$56 \div 7 = 8$

INTRODUCING THE STRATEGY

Guided Conversation

Step One: Select a number to represent a total (a dividend). Also decide on one factor.

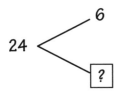

Step Two: Figure out what the second factor should be.

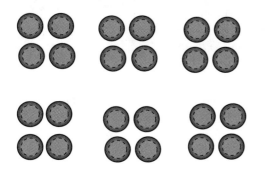

Step Three: Find other ways to break up the total into two factors. Are there only two factors? How many can you find?

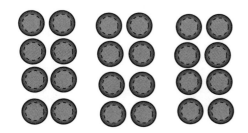

Step One: Today we're going to work on using number bonds to help us learn about division. We're going to start by finding a missing factor. We'll know the total and one of the factors. Do you remember doing number bonds when we were studying multiplication? If so, this is going to look very familiar. I'm going to choose 24 as our total, or dividend, and then 6 as a factor. Look what the bond would look like for this equation.

Step Two: Now get out 24 counters. We said one factor was 6, so let's put our counters into 6 equal groups. How many do you think will be in each group? You say 4? Prove it to me. That's right. When we took our 24 counters and put them in 6 groups, we found out that each group had 4 counters in it.

Step Three: Put the 24 back together into one pile. Let's think of other number bonds for the number 24 that will help us with division. Now let's try putting our counters in 3 groups. This is a way of playing with a different division fact: 24 ÷ 3. How many counters will be in each group? Yes, there will be 8. I wonder if there are any other ways that will work. Let's try 8 groups. How many counters will be in each group? You're right; there are three.

Step Four: Introduce division vocabulary terms: dividend, divisor, and quotient. Notice patterns when looking for factors, and share what you discover with your partner, group, or class. Write the number bonds.

Dividend ÷ Divisor = Quotient

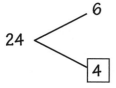

Step Four: Do you remember how to write number bonds? Of course you do. Here's the vocabulary we're going to need to know and use: Dividend ÷ Divisor = Quotient. The dividend is a fancy math word used in division for the total. In multiplication we called it the product. The divisor is the number we are dividing by. It is one of the factors. The quotient is the specific math term for the answer to a division problem. It could also be called a factor, just like in multiplication. Wow! That's a bunch of new vocabulary!

We're going to start with the dividend 24. What is one way I can divide 24 evenly? Use your counters if you need to. Yes, in 6 groups, and 6 is the divisor. How many will be in each group? That's correct; 4, and 4 is our quotient.

Building On

Exploring division through the use of number bonds works with numbers with greater values, decimals, and even fractions.

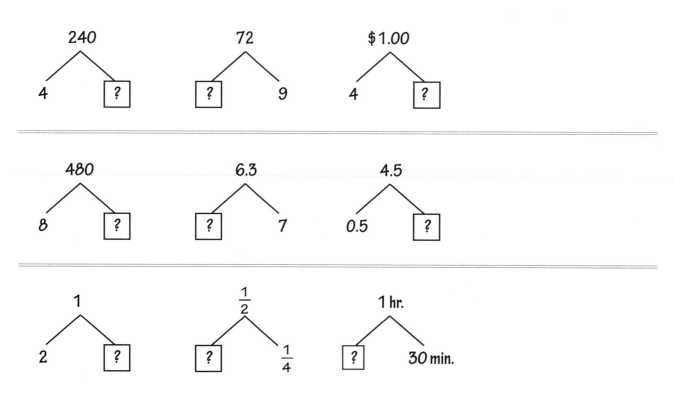

Place Value Disks and Charts

AFTER LEARNING THE BASIC CONCEPT OF DIVISION through practice with manipulatives and number bonds, it is time to introduce division of greater-value numbers using place value disks and charts. This strategy emphasizes the importance of the concrete approach to building comprehension. Students are able to manipulate the disks to show equal sharing. Be sure to emphasize regrouping or trading, why it happens, and how it is recorded on paper. After practicing division with disks, you'll want to have your students draw what's happening. Remember that drawing comes before the abstract computation. Build understanding!

One of the most challenging aspects of teaching long division is making sure that students understand what they are doing and why. Division is complex because the process also involves multiplication and subtraction! At first, guide your students to practice long division using place value disks without recording anything on paper. After students practice with disks only, you can coach them to begin recording what they are doing. The key is for students to understand that, at every step, they need to notate in their abstract recording what they have just done with their disks.

In the following discussions, I've started out with a guided conversation for a lesson in which students manipulate place value disks (without recording) in order to solve a division problem. After that, you'll find a conversation for a lesson in which students will draw place value disks and record their work step-by-step to find the answer.

INTRODUCING THE STRATEGY
Working with disks only

Step One: Select a division problem. Use place value disks to build the dividend.

$$62 \div 4$$

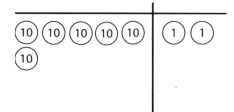

Step Two: Identify the divisor. Draw that number of rows on the place value chart.

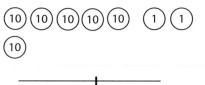

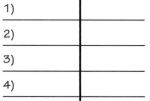

Step One: Write the division expression 62 ÷ 4 on your paper. Think about what it means. Who can tell us what it means? It means there are 62 objects that we want to put in groups of 4. Or it also can mean that we want to put 62 objects into 4 groups. I think the second idea of putting 62 into 4 groups with our disks sounds easier to do. What do you think? Who can tell us a word problem that matches this expression? Let's see if I got all the details of that problem. There were 62 pencils in a box. The pencils were shared equally by 4 students. How many pencils did each student receive? Good story!

Now let's find the answer. First, we need to build the number 62 with disks. How do we do that? You're right. We need 6 tens and 2 ones. Everyone get out your disks and arrange them on your place value chart. Remember to put them in rows of five like a ten frame.

Step Two: The number we're dividing by is called the divisor. It tells us how many groups we need. What are we dividing 62 by? That's right; by 4. We're going to draw 4 rows on our place value chart so that we can keep our disks organized. It will also help us remember that we're dividing the 62 into 4 equal groups. I'm going to draw 4 horizontal sections on my place value chart. Why did I draw 4 horizontal lines? That's right; because we're dividing 62 into 4 groups. Now you draw horizontal lines on your place value chart so it looks like mine. I'm going to number each row so I know that I have set up my chart correctly.

Step Three: Starting with the largest place value, distribute the disks equally among the rows.

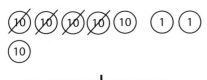

1)	(10)	
2)	(10)	
3)	(10)	
4)	(10)	

Step Four: If there are remaining hundreds or tens disks that cannot be distributed equally, trade them for an equivalent number of lesser-value disks.

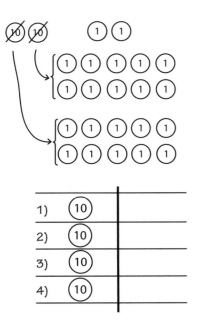

1)	(10)	
2)	(10)	
3)	(10)	
4)	(10)	

Step Three: Let's begin with the largest place value. What disk has the largest value? That's correct; the tens. Do we have enough tens to put one in each of our rows? Yes. Go ahead and place them in each row. This reminds me of when we pass out things in our classroom. One for you, one for you, one for you.

How many tens did you put in each row? 1. Should we give each row another ten? No. Why not? There are only 2 tens left. That's not enough for all 4 groups.

Step Four: Now let's look at what we have left. We have 2 tens and 2 ones. Do we have enough tens to share them equally in the 4 rows? We do not, so we need to make a trade just like we do in subtraction. What are we going to trade the tens disks for? We're going to trade them for ones. We have 2 tens disks left. How many ones will we trade the 2 tens disks for? Yes, we'll get 20 ones disks, because 2 tens equal 20 ones.

Step Five: Divide the lesser-value disks equally among the rows on the chart in the appropriate column. Continue until there are no more disks to distribute.

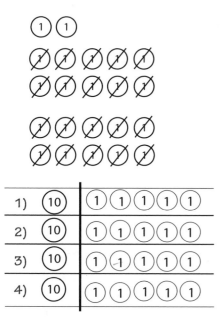

Step Five: Now we don't have any tens disks left, but we have a lot of ones. We need to divide the ones disks equally among the 4 rows. Should we do it one at a time, or can we do more than one for each group? What would you like to try? Okay, we can try 2 for each. Do we have enough? Yes. Do we have enough left to share them again? Yes, we do. How many do you want to give this time? How about 3 for each? Let's try it. Wow! It worked. Wait a minute. I just noticed something. We used 20 disks to put into 4 rows. First, we put 2 in each row and then 3 more in each row. That's a total of 5 ones in each row. 4 x 5 = 20. Maybe next time I'll think about my multiplication facts to help me. Do we have enough to share more? No, we have 2 left, but that's not enough to share equally.

Step Six: Count the number of disks in each row to make sure they are equal. Record any remaining disks as a remainder.

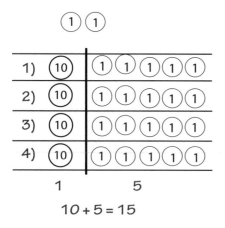

1

5

10 + 5 = 15

62 ÷ 4 = 15 r2

Step Six: Now all we have left to do is count the number of disks in each row and see how many disks we have left as a remainder. Then we'll have our answer to 62 ÷ 4. Count the number of the disks in each row to make sure each row has the same total. We should have 1 ten and 5 ones in each row. 1 ten and 5 ones is the same as 15. What about the 2 remaining ones? We can record them as a remainder. So our quotient would be 15 r2.

Building On

When you provide division problems for students to solve using place value disks, make sure you select numbers that are reasonable to work with. The examples that follow use two-digit and three-digit numbers only. This ensures that kids are able to learn the process without getting bogged down with too many disks. It's certainly possible to use place value disks to solve division problems involving four- and five-digit numbers, but you can hope that this won't be necessary. If students do not understand the concept of place value division with two- and three-digit numbers, then they should not move on to higher-value numbers. Instead, have them review and build the division concept.

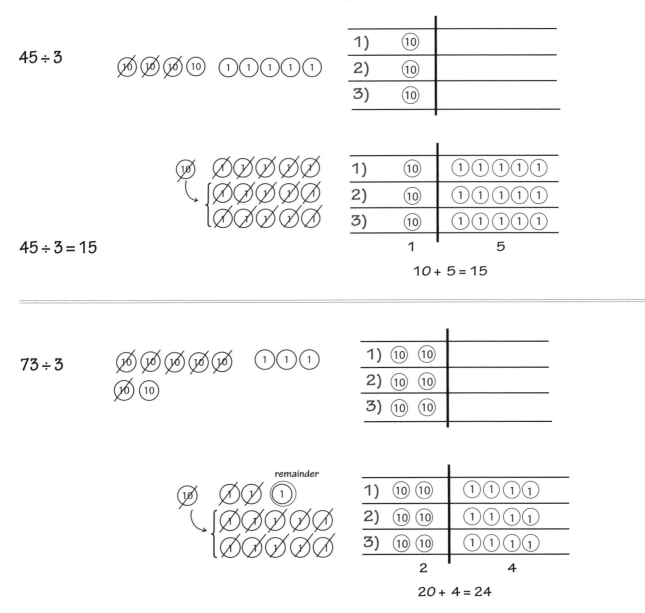

$45 \div 3$

$45 \div 3 = 15$

$10 + 5 = 15$

$73 \div 3$

remainder

$20 + 4 = 24$

$73 \div 3 = 24\,r1$

$633 \div 5$

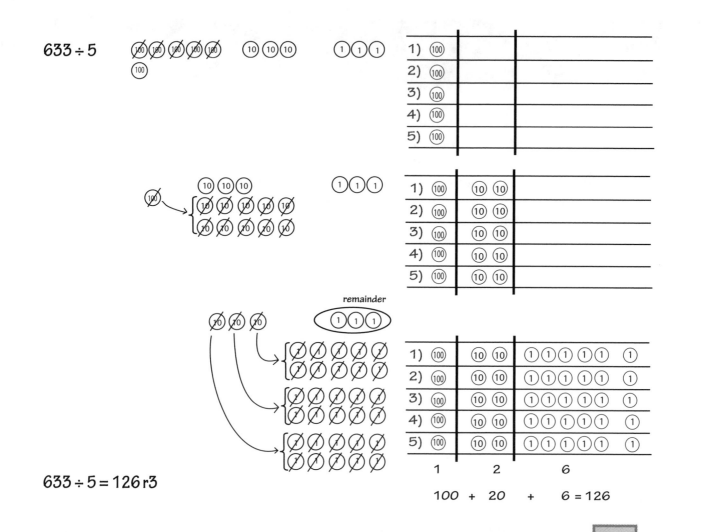

$633 \div 5 = 126\,r3$

	1	2	6
	100	+ 20	+ 6 = 126

Building On and On

You can use place value disks to solve division problems involving decimals too, as long as you have decimal tiles available for your students.

$41.6 \div 4$

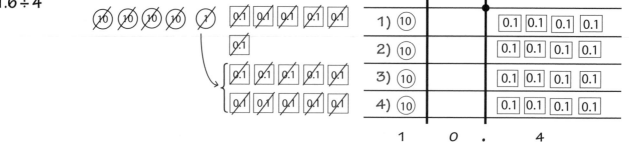

	1	0	.	4

$$10 + 0 + 0.4 = 10.4$$

$41.6 \div 4 = 10.4$

Step One: Select a division problem and write it in traditional form. On paper, draw place value disks to represent the dividend.

$$3\overline{)134}$$

○　○○○　○○○○

Step Two: Next to or below the drawing of your place value disks, draw a place value chart. The number of columns depends on the number of places in the number you are using. Next, identify the divisor. Draw that number of rows on the place value chart.

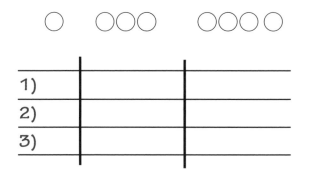

Guided Conversation

Step One: We've been working with place value disks to show what happens when we solve a division problem. Today we're going to take this strategy a step further. To start with, we'll write the division expression $134 \div 3$, and we're going to write it so that it looks like long division. What does this expression mean? You're right, it means we have 134 objects and we want to put them into 3 equal groups.

Now, instead of using place value disks, we're going to draw the disks on paper. Then we will visualize what we would do if we actually were working with real disks. We are also going to record what we're doing one step at a time. Who remembers what we did first when we used place value disks for solving a division problem? We built the dividend. What's our dividend? 134. How do we draw 134? That's right; we draw 1 hundreds disk, 3 tens disks, and 4 ones disks. Double-check with a neighbor that your drawing is correct. Remember to show the disks arranged by place value.

Step Two: What's our next step? Think back to when we had disks out on our desks. I'm remembering that we need to begin with three columns on our place value chart: one each for the hundreds, tens, and ones. Next, we need to look at our divisor, which is 3, and draw 3 rows on our place value chart. Why do we need 3 rows? Because we're dividing by 3. After we draw our lines, we're going to number each row.

Step Three: Starting with the largest place value, draw disks, distributing the disks equally among the rows. If necessary, show a trade of 1 hundreds disk for 10 tens disks.

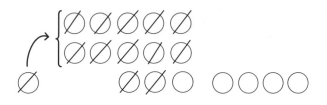

1)	◯◯◯◯	
2)	◯◯◯◯	
3)	◯◯◯◯	

Step Four: Record your work.

$$\begin{array}{r} 4 \\ 3\overline{)134} \\ 120 \\ \hline -\ 14 \end{array}$$

Step Three: Visualize what we did with disks. We began with the largest place value, and in this case that's the hundreds. Do we have enough hundreds to put one in each row? No. What are we going to do instead? We're going to trade in the hundreds disk for 10 tens. Go ahead and draw those 10 tens. What do we need to do with our hundred? Yes. We should cross it out because we don't have it anymore. We have 10 tens instead of the hundred. How many tens do we have altogether? We have a total of 13 tens now. Check to make sure that's what you've drawn.

Now that we've traded in the hundred for tens, we need to divide the tens into 3 equal groups. Do we have enough tens to put 1 in each row? Yes. What are you telling me? We have enough to put more than 1 in each row? Wow! Go ahead and see how many tens you can put in each row. Remember that each row must have an equal number of tens. How many tens did you put in each row? 4. How many tens are left? Only 1.

Step Four: Now we need to record what we've done. Look at your division expression. After we traded the hundred for tens, how many tens did we have? 13. We then put them into equal groups. How many tens did we put in each group? 4. Since we've figured out we need 4 tens, we need to write the 4 above the division symbol in the tens place. What is the value of the tens in each row? There are 4 tens, or 40, in each row. How many rows of 4 tens did we use? 3. That means we used 3 rows of 4 tens, and 3 rows of 40 equals 120. Watch how I record that. I write 120 under the 134. I need to find out how much I have left. How do you think I'll do that? That's right. I'll subtract 120 from 134. That equals 14. Does that match your picture? You should have 1 ten and 4 ones left in your drawing. Notice how what you've written matches your drawing.

Step Five: If there are remaining hundreds or tens disks that cannot be distributed equally, trade them for lesser-value disks. Continue until there are no more disks to distribute. At each step, also record your work.

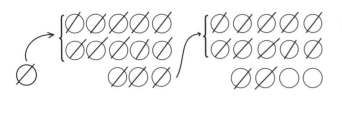

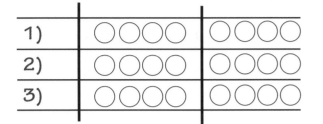

Step Six: At each step record your work. Record any remaining disks as a remainder.

$$3 \overline{)134} \quad 44\,r2$$
$$120$$
$$-\,14$$
$$-\,12$$
$$2$$

Step Five: What do you think we should do next? We need to look at what we have left. What do we have left? We have 1 ten and 4 ones. Do we have enough tens to share them equally among the 3 rows? No. What do we need to do? You're right; we need to trade in the ten for 10 ones. Let's draw this trade. Visualize or act out what you would have done with disks, and draw the result instead. So cross out the tens disk and draw 10 additional ones.

We don't have any tens disks left, but we have some ones. How many ones do we have? 14. Do we have enough ones to share them equally among 3 rows? Can we do it more than one at a time? Do what makes sense to you. Remember to cross out the ones disks that you use as you draw them in a row on the place value chart. Make sure the number of ones disks is equal in each row. How many ones disks did you draw in each row? 4. What is the value of these? 4. We used 3 rows of 4 disks, which equals 12. Do we have any ones left? Yes, we have 2. Is that enough to share? No, so we'll call the 2 leftover disks the remainder.

Step Six: We've finished drawing this stage of our work, so now it's time to record what we've done in the problem. Look at what we already have, and think about what we just did. We traded in the 1 ten for 10 ones, which gave us 14 ones altogether. We drew 4 ones in each row. So we need to record those 4 ones above the symbol next to the 4 tens. We drew 4 disks in each of the 3 rows. How many total ones disks was that? 12. That means we used 12 ones disks. Write the 12 we used below the 14, which was what we had left after we worked the hundreds and tens. Now we need to figure out again how much we have left. We'll do the same thing we did earlier. We'll subtract. This time we'll subtract 12 from 14. How many are left? 2. Does that match our disk drawing? Yes. We'll write the 2 leftover disks as our remainder.

Look at what we've written as our answer. We've written 44 with a remainder of 2. Double-check that each row is equal and that our writing matches our drawing. Does our writing match our drawing? Yes. So, 134 ÷ 3 = 44 r2.

Repeated practice with pictorial representation is an important part of the process for building comprehension of division concepts. I would suggest working through several pictorial representations and connecting them to the abstract as an entire class. This gives kids an opportunity to explain what they're doing and why while using the correct language. Choose numbers that do not require too much time spent drawing. This exercise is about building conceptual understanding, not drawing forever. As you and your students practice, make sure they draw, then record, at each step in the process. One step at a time, with partner work for language practice, is the key!

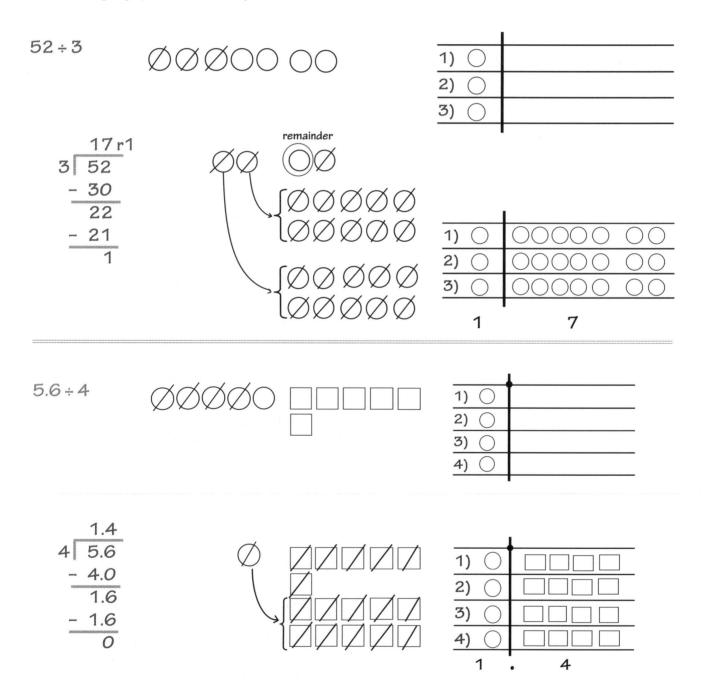

455 ÷ 4

1) ◯		
2) ◯		
3) ◯		
4) ◯		

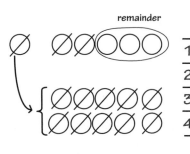

1) ◯	◯	
2) ◯	◯	
3) ◯	◯	
4) ◯	◯	

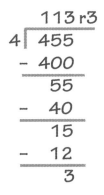

$$\begin{array}{r} 113\ r3 \\ 4\,\overline{\smash{)}\,455} \\ -\ 400 \\ \hline 55 \\ -\ 40 \\ \hline 15 \\ -\ 12 \\ \hline 3 \end{array}$$

remainder

1) ◯	◯	◯◯◯
2) ◯	◯	◯◯◯
3) ◯	◯	◯◯◯
4) ◯	◯	◯◯◯
1	**1**	**3**

The Distributive Property

AFTER TEACHING DIVISION AS PUTTING OBJECTS **into equal groups**, emphasizing number bonds, and using place value disks and strips to build understanding for division of multidigit numbers, it is time to stretch this learning to the distributive property.

The distributive property for division is slightly different from that for addition and multiplication. In both addition and multiplication, the distributive property and left-to-right strategies generally focused on decomposing numbers by place value. This is not necessarily the case for division. The distributive property for division focuses on breaking the dividend (total) into friendly numbers. When the topic is division, the term friendly number refers to any number that can be easily and evenly divided by a one-digit number. For example, with the expression $348 \div 3$, 348 might become $300 + 48$, because $300 \div 3$ and $48 \div 3$ are easy to divide evenly. Or, if a student didn't know the answer to $48 \div 3$, he could decompose 48 into 30 and 18. Then the student could work with the original $300 \div 3$, along with $30 \div 3$ and $18 \div 3$, which would result in $100 + 10 + 6 = 116$.

You may be thinking, "Wow! That takes a lot of thinking and math sense." You're right. But if we've done our job laying the groundwork by teaching place value and number bonds, then our kids are ready.

You may also notice that your students will solve a division problem such as $348 \div 3$ by decomposing it in different ways. For example, because 48 can be decomposed into 24 and 24, another approach to $348 \div 3$ is this: $300 \div 3 = 100$, plus $24 \div 3 = 8$, plus $24 \div 3 = 8$. Adding these partial quotients, $100 + 8 + 8$, of course produces the same final quotient: 116.

One of the keys to success with different strategies is knowing which strategy is most efficient. The distributive property strategy is efficient for friendly numbers. If you need to solve problems with numbers that are not friendly, you should use the partial quotient strategy instead. We will explore the partial quotient strategy in the next section of this chapter.

The following strategy and guided conversation introduce how to divide a three-digit number by a one-digit number using the distributive property. Keep in mind that while I present this example using abstract thinking, it could easily be modeled using place value disks or strips to make it concrete.

INTRODUCING THE STRATEGY

Step One: Write the division expression horizontally.

$345 \div 3$

Step Two: Look at the dividend (total) and check carefully for ways it could be decomposed to make division easier. Think: "Can I break the number into friendly numbers that are evenly divisible by one divisor?" Write down the parts. If your first try includes numbers that aren't friendly numbers for dividing by 3, try again and decompose the total in a different way.

$345 = 300 + \cancel{40} + \cancel{5}$

                unfriendly

$345 = 300 + 30 + 15$

Guided Conversation

Step One: Look at the division expression I've written on the board. What do you notice? It's a three-digit division expression. We haven't done three-digit division before, but I know you can do it. Be ready to use all of the number bond ideas we've learned.

What does $345 \div 3$ mean? That's right; it means we have 345 objects and we are splitting them into 3 equal groups. We want to know how many are in each group. What else could it mean? It could mean we have 345 objects that we are putting in groups of 3, and we want to know how many groups of 3 there will be. Can you visualize that? That would be a lot of groups!

Step Two: I'll start by looking at the dividend. Which number is the dividend? That's correct; it's 345. What are we dividing by? That's right; by 3. Our challenge is to decompose 345 into parts that are easier to divide by 3 than the entire 345. What is one way we could decompose 345? We could decompose it into $300 + 40 + 5$. Let's check whether each part is easy to divide by 3. Can we easily do $300 \div 3$? Yes. What about $40 \div 3$? No, it would have remainders. So, we need to find a different way to decompose 345. Who has a different idea? Since we said $300 \div 3$ was easy, we need to concentrate on the 45. How could we decompose 45 into friendly parts that are easy to divide by 3? How about 30 and 15? Are both of these numbers easy to divide by 3? Yes. So we'll decompose 345 into $300 + 30 + 15$.

Step Three: Divide each of the parts separately by the divisor.

$345 \div 3$

$(300 \div 3) + (30 \div 3) + (15 \div 3)$

$\quad 100 \quad + \quad 10 \quad + \quad 5$

Step Four: Add the partial quotients together to get your final answer.

$345 \div 3$

$(300 \div 3) + (30 \div 3) + (15 \div 3)$

$\quad 100 \quad + \quad 10 \quad + \quad 5 \quad = 115$

$345 \div 3 = 115$

Step Three: Now it's time for us to divide each of the parts. We'll start with the largest place value, which is the hundreds. How many hundreds? 3. What is its value? 300. Write down $300 \div 3$. How much is $300 \div 3$? Yes, it is 100. What's our next part? 30. Why? Because the tens are the next-largest place value. We need to write down and compute $30 \div 3$. The answer is 10. Now we need to divide our last part, $15 \div 3 = 5$. That's a simple number bond.

Step Four: We're almost done finding the quotient. All we have to do is put the parts back together. Why do we have to add these numbers? That's right. We began by pulling the numbers apart, or decomposing them. Now we need to put them back together, and we do that by adding. What parts do we have? We have 100, 10, and 5. Let's write them in an addition expression and find the total. $100 + 10 + 5 = 115$. We have just found the quotient for $345 \div 3$. It is 115.

Here are more examples of solving division expressions using the distributive property. You'll see that this method can be used with multidigit numbers as well as decimals. Remember, there are often several different ways to decompose the numbers in any given expression. As you work through these examples, ask your students to find another way to solve each expression.

$57 \div 3$

$(30 \div 3) + (27 \div 3)$

$\quad 10 \quad + \quad 9 \quad\quad = 19$

$624 \div 4$

$(400 \div 4) + (200 \div 4) + (24 \div 4)$

$\quad 100 \quad + \quad 50 \quad + \quad 6 \quad = 156$

$3,642 \div 6$

$(3,000 \div 6) + (600 \div 6) + (42 \div 6)$

$\quad 500 \quad + \quad 100 \quad + \quad 7 \quad = 607$

$27.9 \div 9$

$(27 \div 9) + (0.9 \div 9)$

$\quad 3 \quad + \quad 0.1 \quad = 3.1$

$68.4 \div 9$

$(63 \div 9) + (5.4 \div 9)$

$\quad 7 \quad + \quad 0.6 \quad = 7.6$

Partial Quotient Division

DIVISION USING THE PARTIAL QUOTIENT STRATEGY is a great alternative to long division for students who have not yet mastered all of their basic division facts. It's a fabulous strategy for showing what is happening with more involved division problems one stage at a time. Also, it's a strategy that can be used for teaching division problems with and without remainders.

The partial quotient strategy reminds me of what often happens in elementary classrooms when a student has a bag of candy to share and has no idea of how many pieces to give each person in order to share the goodies equally all around. So, she begins by handing out two pieces each, then decides there is enough remaining to hand out another two each. This continues as she keeps checking to see whether there's enough candy left in the bag to hand out even more. The student keeps dividing the total by parts until she can no longer share the remainder fairly.

As you look through the examples of the partial quotient strategy on the following pages, you'll see that it allows students to use their multiplication facts (number bonds) to help solve division problems. The beauty of this strategy is that students can use the multiplication facts that they already know to help them. For example, a student who may not have mastered his multiplication or division facts for 8 is able to use what he knows about multiplication and division facts for 5 and 3 to find a final quotient.

INTRODUCING THE STRATEGY

Step One: Select a division problem. Set it up like this.

Step Two: Think about how many groups of 3 can be made from the dividend. Select a number and try it. Write your number on the right side of the line. Multiply your guess by the divisor. Subtract this product from the total you began with.

$$3\overline{)576} \\ \underline{-300} \qquad 100 \\ 276$$

Step One: We are going to use what we know about number bonds and the multiplication and division facts we are experts with to solve this multidigit division problem: 576 ÷ 3. Wow! Do you think we can do it? Of course we can. We're also going to use what we already know about multiplication and place value to help us. Write your expression like this. Notice that I wrote the dividend inside the division symbol and that I am making a long line down the right-hand side. Then I'll write the divisor to the left. Look at what I've written and make sure what you've written looks the same.

Step Two: Start by thinking about how many 3s could be in 576. Pick a number that's easy to mentally multiply. What would you like to try? Okay, let's start with 100. Why did you choose 100? Because 3 x 100 is easy to do. Write the 100 to the right of the line. Now we need to multiply. What is 3 x 100? 300. Write the 300 below the 576 and subtract. What is 576 – 300? 276. Look at my example. Make sure your work looks like mine.

What if I had chosen 200? Then I would have multiplied 3 x 200. What would the answer be? Yes, 600. I would notice that 600 has a greater value than 576. I would have had to revise my idea and try a number with a lesser value.

Step Three: Look at how much is left. How many groups of 3 can be made from this number? Record your guess below the first one to the right of the line. Multiply and subtract again. Continue until you can no longer divide.

```
3 | 576
  - 300 | 100
    276
  - 150 |  50
    126
```

```
3 | 576
  - 300 | 100
    276
  - 150 |  50
    126
  - 120 |  40
      6
```

```
3 | 576
  - 300 | 100
    276
  - 150 |  50
    126
  - 120 |  40
      6
  -   6 |   2
      0
```

Step Three: How much do we have left now? That's right; we have 276 remaining. Do we have enough left to try another 100? No, because we have less than 300. So, what should we try? How about 50? Why do you suppose we chose 50? Because 3 x 50 = 150, and 150 is less than 276. Write 50 to the right of the line. Now it's time to multiply. What is 3 x 50? 150. Write 150 below the 276 and subtract. What is 276 – 150? 126. Do you think we could have chosen a number other than 50? Of course we could have. What could we have chosen? 60? Why? We could have chosen 60 because 3 x 60 = 180 and that's less than 276. 80? Why or why not? We could also have chosen 80 because 3 x 80 = 240, which is less than 276. This time, though, we are going to stick with 50.

How much do we have left now? We have 126 left. How many 3s should we try now? We can't choose 50 again. Why? Because 3 x 50 = 150, and we have only 126 left. How about 40? Will it work? Why? It'll work because 3 x 40 = 120, and 120 is less than 126. Write 40 to the right of the line. Multiply 3 x 40. What's the product? Yes, it's 120. Write 120 below 126 and subtract. That leaves 6.

We have 6 left. How many 3s in 6? That's easy; there are 2. Multiply 3 x 2. Write the 6 below what's left in our work and subtract. What's left? Zero. We're done dividing. How do I know that we're done dividing? Because there is nothing left.

Step Four: Add up the partial quotients on the right-hand side of the vertical line to determine your final quotient.

$$100 + 50 + 40 + 2 = 192$$
$$576 \div 3 = 192$$

Step Four: Our final step is to add the partial quotients that we've written to the right of our line. When we divided, what parts did we figure out? We figured out 100, then 50, 40, and finally 2. What is 100 + 50 + 40 + 2? You're right. It's 192. Now we know the answer to 576 ÷ 3. Pretty cool, huh! I love how we used multiplication facts that we already know to solve this long-division problem.

Building On

The partial quotient method opens up many different ways to solve a single division problem. Here, only one solution is shown for each problem, but you should encourage your students to show and share multiple ways to solve these problems using the partial quotient strategy. It even works with division that has remainders. You'll notice that the method is the same. The only difference is that there is a quantity left at the end of the dividing that needs to be written as a remainder.

$79 \div 4$

```
          19 r 3
      4 |  79
       -  40  | 10
          39
       -  20  |  5
          19
       -  16  |  4
           3  |
```

$$10 + 5 + 4 = 19$$

$$79 \div 4 = 19\,r3$$

$652 \div 5$

```
          130 r 2
      5 |  652
       -  500  | 100
          152
       -  150  |  30
            2  |
```

$$100 + 30 = 130$$

$$652 \div 5 = 130\,r2$$

You can use the partial quotient method with higher-value numbers and decimals, too.

$2,456 \div 14$ $78.4 \div 7$

```
              175 r 6                              11.2
         14 | 2,456                            7 | 78.4
          -  1,400 | 100                        -  700 | 100
             1,056                                 84
          -    700 |  50                        -  70 |  10
               356                                 14
          -    280 |  20                        -  14 |   2
                76                                  0
          -     70 |   5
                 6 |
```

$100 + 50 + 20 + 5 = 175$ $100 + 10 + 2 = 112$

Remember to put the
decimal point back in!

$2,456 \div 14 = 175 \, r2$ $78.4 + 7 = 11.2$

Traditional Long Division

TRADITIONAL LONG DIVISION SHOULD BE TAUGHT only after students have a solid understanding of the concept of two- and three-digit division. When teaching the traditional division algorithm, connect what is happening abstractly with earlier work with place value disks and drawings. Have students pretend to manipulate the disks before recording. (Act it out.) Students should be asking themselves: "What's the next step? Do I need to trade? Why or why not?" When you begin traditional long division, I strongly suggest that as students learn the abstract way of writing out their work, they also use place value drawings on the same page. (For an example of a guided conversation on drawings of place value disks, see page 87.) As your students work, have them refer back and forth between the abstract and pictorial and connect the methods. Otherwise, traditional long division becomes a series of rote steps that will not lead to comprehension and long-term retention of concepts.

INTRODUCING THE STRATEGY

Step One: Write a division expression using the vertical symbol for division.

$$3 \overline{)472}$$

Step Two: Starting with the largest place value (in this case, hundreds), consider how many of the divisor there would be in that quantity. Write the number you're thinking of above the hundreds place of the dividend. Remember to use a single-digit number only. Multiply this number by the divisor. Record the product below the hundreds place in the dividend and subtract to find the remainder.

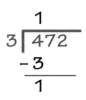

Guided Conversation

Step One: Please write the expression 472 ÷ 3 on your paper using the vertical symbol for division. What's the other way we can write this that means the same thing? That's right; we could write it horizontally. The way we are setting up the problem today looks similar to the partial quotient method, but we will be recording our work differently this time.

Step Two: Now that we've written the problem we're going to start to solve it. I'm thinking about how many 3s can be in 400. The answer is 100. We'll write this answer at the top in the hundreds place. But we're going to use a shortcut. We're going to write a 1 only. Next, since we wrote that the answer is 1 in the hundreds place, we need to multiply 3 x 1. Then we write the answer, 3, under the 4 to show that it represents 3 hundreds. Now we subtract. How many hundreds are left? That's correct; there is 1 hundred left. If we imagine working with disks, what does this 1 that we wrote up above represent? It represents 100, and the 3 that we subtracted represents 300. Notice how this looks very similar to what we did with the partial quotient strategy. What's the same? What's different?

Step Three: Rename the remaining hundred(s) as tens and add them to the tens. Record that total by bringing down the digit in the tens column and rewriting it as part of the remaining tens. Divide these remaining tens by the divisor: come as close as possible to the number of tens without going over. Write that number above the tens place of the dividend. Multiply that number by the divisor. Record the product below the remaining tens and subtract.

```
        1 5
   3 ) 4 7 2
     - 3 ↓
     ─────
        1 7
     -  1 5
       ─────
           2
```

Step Three: We have only 1 hundred left and that can't be divided easily by 3, so we're going to trade in the hundred for 10 tens. Remember how we did this with our disks? When we put together the 10 tens with the 7 tens that are already there, we have a total of 17 tens. We'll show this by "bringing down" the 7 in the tens place next to the 1 remainder. It looks like a 17, but we know it's really 17 tens, or 170. Next, we ask, what is 17 tens divided by 3? We want to get as close to 17 as we can without going over. I know that 5 groups of 3 go into 17, so I'm going to write down a 5 in the tens place above the division symbol. Now we need to multiply 5 x 3 and write the product under the 17. Subtract 17 – 15 to find how many tens remain. What did you get? 2. That's right. But remember, it's in the tens, so it's equal to 20. Think back to when we were using place value disks. See how this almost matches? It's just a little shortcut.

Step Four: Rename the remaining ten(s) as ones and add them to the ones. Record that total by bringing down the digit in the ones column and rewriting it as part of the ones remainder. Divide the ones remainder by the divisor, coming as close as possible to the number of ones without going over. Write that number above the ones place of the dividend. Multiply that number by the divisor. Record the product below the ones remainder and subtract. The number above the division symbol is your quotient. If there is any final remainder, write it next to the quotient, using a lowercase r to represent the remainder.

```
        1 5 7 r1
   3 ) 4 7 2
     - 3 ↓
     ─────
        1 7
     -  1 5 ↓
       ─────
          2 2
      -   2 1
         ─────
              1
```

Step Four: We're headed to the ones now. Can we divide the 2 leftover tens into 3 groups? If we were using disks, we would make a trade, wouldn't we? Why would we do that? Because we couldn't divide the 2 tens disks evenly among the three rows on our place value chart. So, we need to rename the 2 tens as 20 ones and then add the ones that are already there. We record that by bringing down the 2 ones and writing a 2 next to the 2 tens. That gives us 22. Now we need to divide 22 ones by 3 without going over 22. We know that 3 x 7 is 21, so we'll write 7 above the symbol and then write the 21 below the 22. Following the pattern, what do we do next? Subtract 22 – 21. The answer is 1, and since we can't easily divide 1 one by 3, we know that 1 is our remainder.

The traditional method for long division is going to take a while for kids to digest. I strongly recommend that you use the guided release of responsibility model of teaching here. In other words, after modeling the strategy while working through several examples together as a class, have the students work in small groups or with partners to practice the traditional method. After you hear and see kids using the method correctly, then allow them to work independently.

Once you are certain that your students understand this method, you can challenge them with problems involving four- and five-digit numbers or even decimals (if your students are already familiar with decimals). Kids who truly understand the long-division process will be able to apply it to higher-value numbers and decimals with confidence.

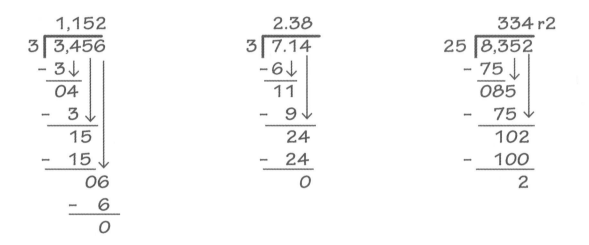

Short Division

SHORT DIVISION IS EXACTLY AS THE NAME IMPLIES . . . short. It is a shorter way, requiring much less writing than long division. Short division is not for everyone. It is for students who are very successful with long division, excellent mental mathematicians, and those who need a new challenge. This method requires students to "skip" the recording of the traditional steps of multiplying and subtracting. Instead, they'll multiply and subtract mentally, as you'll see in the examples that follow. The only steps recorded are the quotient and the remainders that result at each stage in solving the equation.

As students begin this method, remind them to think about long division and the steps they followed. They'll be doing exactly the same thing, but instead of writing down the product each time and subtracting to get a remainder, they'll be doing their figuring mentally. In fact, I suggest first using long division to solve a division problem and then writing the short-division version right next to it so that kids can see the connection between the two.

You may be saying to yourself, "Will this really work? What about division with two-digit divisors?" To this I say, try it, you'll like it, and so will your students.

INTRODUCING THE STRATEGY

Step One: Select a division expression and write it down traditionally.

Step Two: Figure out how many times the divisor can go into the digit with the largest value in the dividend. Multiply this number by the divisor to get a partial quotient. Mentally subtract, and then write the remainder.

Step Three: Figure out how many times the divisor can go into the second-largest digit. Multiply this number by the divisor to get a partial quotient. Mentally subtract, and then write the remainder.

Guided Conversation

Step One: You kids are getting so good at long division that I thought I'd show you a new way that requires a lot less writing. Sounds great, doesn't it? The trick is that you will have to do a lot of mental math instead. Our first expression is 556 ÷ 3. Let's write it the same way we would if we were doing long division.

Step Two: We'll begin by thinking about how many 3s go into 5. How many? 1. Record the 1 above the division symbol like we always do. What is 3 x 1? 3. Here's where the tricky mental work begins. Mentally put the 3 below the 5 and subtract. How much is left? That's right; 2. Look carefully where I'm going to put the 2. I put it right up by the 5 tens so it looks like a 25. Notice what I'm doing: I'm writing the remainder up above and ahead of the next digit in the dividend.

Step Three: Now we're going to move to the next-largest place. We need to think how many 3s go into 25. You say 9? Remember that 3 x 9 is 27, so that's a little bit too big. How about 8? I know that 3 x 8 is 24 and that's close to 25. Write the 8 up above the division symbol where the quotient (answer) goes. Mentally put the 24 you just got from multiplying below the 25 and subtract. How much is left? 1. We're going to put the remaining 1 up by the 6 ones so it looks like a 16. Check your recording to make sure it matches mine.

Step Four: Continue the process until it's not possible to divide any longer.

$$\begin{array}{r} 185\,r1 \\ 3\,\overline{)\,5\overset{2}{5}\overset{1}{6}} \end{array}$$

Step Four: What do you think we're going to do next? We need to keep going. We're getting close. How many 3s are in 16? If you said 5, you are correct. Write the 5 above the division symbol in the ones place. What is 3 x 5? 15. Mentally put the 15 below the 16 and subtract. How much is left? Only 1. This 1 will be our remainder because we can't make any more groups of 3. I'll write the remainder just like we've done before. Wow! That was a short way to solve a division problem. I guess that's why they call it short division!

Building On

As I stated earlier, short division is not for all students. It is for your more advanced students who need a division challenge. More examples of short division are provided below. As with all computation, this method can also be used with decimals by simply placing the decimal point in the correct position.

$$98.7 \div 4 \qquad \begin{array}{r} 24.6\,r3 \\ 4\,\overline{)\,9\overset{1}{8}.\overset{2}{7}} \end{array}$$

$$2{,}456 \div 4 \qquad \begin{array}{r} 614 \\ 4\,\overline{)\,2{,}4\overset{1}{5}6} \end{array}$$

$$113{,}568 \div 9 \qquad \begin{array}{r} 12{,}618\,r6 \\ 9\,\overline{)\,1\overset{2}{1}\overset{5}{3}{,}\overset{1}{5}\overset{7}{6}8} \end{array}$$

Reproducible Word-Wall Words

array

associative property

commutative property

add

addition

area model

digit

distributive property

divide

decimals

decompose

difference

equal

equation

equivalent

dividend

division

divisor

multiple

multiplication

multiply

expanded
notation

factor

horizontal

partial quotient

place value

product

number bond

part-whole

partial product

sum

vertical

quotient

subtract

subtraction

References

Bruner, J. 2000. *The Process of Education*. Cambridge, MA: Harvard University Press.

Chen, S. 2008. *The Parent Connection for Singapore Math*. Peterborough, NH: Crystal Springs Books.

Dienes, Z. 1971. *Building Up Mathematics*. London: Hutchison Educational Limited.

Gibbons, P. 2002. *Scaffolding Language, Scaffolding Learning*. Portsmouth, NH: Heinemann.

Johnston, P.H. 2004. *Choice Words*. Portland, ME: Stenhouse Publishers.

Keene, E.O., and S. Zimmermann. 1997. *Mosaic of Thought*. Portsmouth, NH: Heinemann.

Kuhns, C. 2006. *Number Wonders*. Peterborough, NH: Crystal Springs Books.

Kuhns, C. 2009. *Building Number Sense*. Peterborough, NH: Crystal Springs Books.

Lee, P.Y. (ed.). 2007. *Teaching Primary School Mathematics*. Singapore: McGraw-Hill.

Leinwand, S. 2009. *Accessible Mathematics*. Portsmouth, NH: Heinemann.

Ma, L. 1999. *Knowing and Teaching Elementary Mathematics*. Mahwah, NJ: Lawrence Erlbaum.

Ng Chye Huat, J. (Mrs.), and Mrs. Lim Kian Huat. 2001. *A Handbook for Mathematics Teachers in Primary Schools of Singapore*. Singapore: Federal Publications—Times Media Private Limited.

Parker, T., and S. Baldridge. 2003. *Elementary Mathematics for Teachers*. Bloomington, IN: Sefton-Ash Publishing.

Skemp, R. 2002. *Mathematics in the Primary School*. London: Routledge.

Van de Walle, J.A., K.S. Karp, and J.M. Bay-Williams. 2010. *Elementary and Middle School Mathematics: Teaching Developmentally*, 7th edition. San Francisco: Allyn & Bacon.

Vygotsky, L. 1978. *Mind in Society: The Development of Higher Psychological Processes*. Cambridge, MA: Harvard University Press.

Wegerif, R., and N. Mercer. 1996. "Computers and Reasoning Through Talk in the Classroom." *Language and Education* 10 (1): 47–64.

Index

A

Abstract thinking or recording, in C-P-A approach, 3, 4, 5
    for addition, 12
    for division, 76, 81, 92, 93–94, 101
    for multiplication, 51, 52, 57, 73
    for subtraction, 36, 46, 48
Acting methods
    as bridge from concrete to abstract, 33
    for demonstrating *horizontal,* 23
    for demonstrating *vertical,* 27
    for showing number bonds, 15, 55
Addition, 11–12
    part-whole concept in, 11, 15, 16, 53
    place value in, 19, 22, 23, 30
    regrouping in, 11–12, 17, 24, 26
    repeated, multiplication as, 51, 76
    strategies for
        decomposing numbers, 11, 12, 19–21
        left-to-right addition, 6, 11, 12, 22–25
        number bonds, 11, 12, 13–18
        place value disks and charts, 11, 12, 26–29
        traditional addition, 11, 12, 33–35
        vertical addition, 11, 12, 30–32
Area model, for multiplication, 51, 52, 67
    Building On, 72
    introducing the strategy, 68–71
Arrays, in area model strategy, 67, 68, 69, 70, 71
Associative property, 67

B

Base ten blocks, for decomposing numbers, 19, 20
Base ten proportionality, 3
Bloom's Taxonomy, 6
Bruner, Jerome, 3
Building On sections. *See also Building On sections in specific strategies*
    purpose of, 8

C

Commutative property, 58, 60, 66
Comprehension, 1, 2, 7
Computation strategies, comparing and contrasting, 7, 11, 30, 31, 33, 51–52, 101
Concrete-pictorial-abstract (C-P-A) approach. *See* C-P-A approach
Core curriculum, Singapore Math as, 2
Counters

    for learning division, 76, 79
    for showing number bonds, 14, 38, 39, 51, 54, 55
C-P-A approach, 1, 3–5
    for addition, 12
    for division, 76, 81
    for multiplication, 51, 52
    for subtraction, 36, 48

D

Decimals
    in addition problems, 16, 18, 21, 25, 32
    in division problems, 80, 86, 95, 100, 104, 107
    in multiplication problems, 56, 62, 63, 64, 66, 72, 73
    in subtraction problems, 40, 46
Decimal strips, 8
Decimal tiles, 8. *See also* Place value tiles, for showing decimals
Decomposing numbers
    for addition, 11, 12, 14, 17, 19, 22, 23
        Building On, 21
        introducing the strategy, 20–21
    for division, 92, 93, 95
    for multiplication, 51, 68, 70
Dienes, Zoltan, 3
Distributive property
    area model and, 67
    for division, 77, 92
        Building On, 95
        introducing the strategy, 93–94
    for multiplication, 51, 52, 63
        Building On, 66
        introducing the strategy, 64–65
        part-whole concept and, 11
Dividend, 78, 80
Division, 76
    regrouping in, 81, 88, 89
    remainders in, 84, 85, 89, 90, 91, 96, 99, 100, 103, 104, 105, 106, 107
    strategies for
        distributive property, 77, 92–95
        number bonds, 77, 78–80
        partial quotient division, 77, 96–100
        place value disks and charts, 77, 81–91
        short division, 77, 105–7
        traditional long division, 77, 101–4
Divisor, 78, 80
Drawing pictures. *See* Picture drawing

Standards Addressed by Why Before How

- Understands the meaning of numbers, relationships among numbers, and different ways of representing numbers. (1–3)

- Reads and writes multidigit whole numbers using base ten numerals, number names, and expanded form. (4)

- Understands the place value system and uses that understanding in performing addition and subtraction problems. (1–3)

- Uses mathematical models to represent the addition and subtraction of whole numbers. (1–3)

- Understands and applies the properties of operations to addition and subtraction. (1–3)

- Uses place value understanding and the commutative and distributive properties to perform multidigit arithmetic. (3–4)

- Demonstrates ability to use multiplication and division to solve problems involving numbers up to 100. (3)

- Understand properties of multiplication and the relationship between multiplication and division. (3)

- Performs calculations with multidigit whole numbers and with decimals to hundredths. (5)

- Works with equal groups of objects to gain foundations for multiplication. (2

- Fluently computes with multidigit numbers and finds common factors and multiples. (6)

- Applies the commutative property and distributive property strategies to multiply and divide. (3–5)

- Understands division as problems involving an unknown factor. (3)

- Fluently adds and subtracts multidigit whole numbers using the standard algorithm. (4)

- Illustrates and explains calculation by using equations, rectangular arrays, and/or area models. (3–4)

- Demonstrates an ability to compute fluently. (1–6)